D1211484

BIOMONITORING AIR POLLUTANTS WITH PLANTS

POLLUTION MONITORING SERIES

Advisory Editor: Professor Kenneth Mellanby

Monks Wood Experimental Station,
Abbots Ripton, Huntingdon

BIOMONITORING AIR POLLUTANTS WITH PLANTS

WILLIAM J. MANNING
Associate Professor, Department of Plant Pathology, University of Massachusetts, Fernald Hall, Amherst, Massachusetts 01003, USA

and

WILLIAM A. FEDER
Professor, Suburban Experiment Station, University of Massachusetts, Waltham, Massachusetts 02154, USA

APPLIED SCIENCE PUBLISHERS LTD
LONDON

APPLIED SCIENCE PUBLISHERS LTD
RIPPLE ROAD, BARKING, ESSEX, ENGLAND

British Library Cataloguing in Publication Data

Manning, William J
Biomonitoring air pollutants with plants.—
Biological monitoring series).
1. Air—Pollution—Measurement
2. Plants, Effect of air pollution on
I. Title II. Feder, William A III. Series
614.7′1 TD 890

ISBN 0-85334-916-9

WITH 51 TABLES AND 36 ILLUSTRATIONS

Photosetting by Thomson Press (I) Ltd. New Delhi
Printed in Great Britain by Galliard (Printers) Ltd, Great Yarmouth

Preface

It is well known that certain species or cultivars of higher vascular plants and many lower non-vascular plants, such as mosses and lichens, respond to gaseous and particulate air pollutants in predictable ways and at concentrations much lower than those that elicit responses in humans and animals. Many can be used as indicators of pollutants in air. The responses of others can be related to known concentrations of specific air pollutants. These plants both indicate the presence of air pollutants and can be used to monitor concentrations of the pollutants. These plants are known as 'Biomonitors' and they are the focus of this book.

A wide variety of literature sources must be consulted to obtain information on the use of plants as biomonitors of air pollutants. Our purpose here is to bring together for the first time in one place information on the current state-of-the-art on using plants of all types to biomonitor a number of air pollutants. This information should be of interest to scientists, government officials, industrialists, environmentalists and others concerned with the pollutants that increasingly pervade the air.

The assistance of Mrs Joyce Mieg and Mrs Orene Berg in typing the manuscript is gratefully acknowledged.

William J. Manning
William A. Feder

Contents

CHAPTER 1

Introduction

Our purpose here will be to try to answer two questions: (a) which air pollutants affect plant ecosystems and (b) where do they come from? To answer these questions, however, requires some background knowledge.

I. THE NATURE OF OUR AIR ENVIRONMENT

Before discussing air pollutants, it is first necessary to define air and, further, that portion of the air that affects the existence of living organisms. The air is an ocean of gases which reaches from the earth's surface to outer space. The portion of this atmospheric ocean most important to life is the troposphere, a layer 5–10 miles deep, next to the surface of the earth. This layer of air, plus the oceans and the earth's outer crust, is sometimes called the biosphere. It is only in this very thin, finite layer that all life exists. It is this layer of air and its interface with the surface of the earth that we must consider in order to understand the fate of pollutants in the atmosphere.

The troposphere is made up of many elements of which the most important are nitrogen (N_2) and oxygen (O_2). In terms of parts per million (ppm), the troposphere contains about 800 000 ppm N_2 and about 210 000 ppm O_2; a ratio of about 4:1. It also contains about 300 ppm carbon dioxide (CO_2) and very small amounts of argon, neon, xenon, hydrogen, nitrous oxide, krypton, methane and helium.

Billions of years ago, before green plants evolved, the ratio of oxygen to carbon dioxide in the troposphere was essentially reversed from what it is today. There was little or no available oxygen in the air and organisms which thrived in a low oxygen environment slowly disappeared as the ratio of oxygen: carbon dioxide slowly evolved to where it is today. Thus, oxygen, which now drives most present biological systems, was an air pollutant several million years ago.

Near the outer edge of the troposphere, about 20–30 000 m above the earth, is a dense layer of ozone (O_3). Ozone is a clear, colorless, highly reactive gaseous trivalent form of oxygen. This ozone layer acts as a filter and reduces the amount of ultraviolet radiation from the sun so that it is biologically safe when it reaches the earth's surface. The presence of this naturally formed ozone layer is essential for the survival of present life systems on earth. This ozone layer is not part of our subsequent discussion of air pollutants. Little contact occurs between tropospheric ozone and the atmosphere. Ozone as an air pollutant is photochemically formed from products generated by the combustion of fossil fuels and is restricted to the lower, denser atmosphere nearer the surface of the earth.

Passing through the troposphere towards the stratosphere and into the ionosphere the air thins rapidly as fewer molecules rise to high altitudes. During this same ascent, the air first becomes much colder, dropping below −40 °C in the ozone layer, then rising to about 85 °C at the lower boundary of the stratosphere, cooling down to about −20 °C in the stratosphere and then rising in a linear fashion through the ionosphere to over 200 °C at about 200 000 m above the surface of the earth.

In the habitat of terrestrial biological systems the temperature changes continuously. In the lower, denser, atmosphere air cools with rising altitudes. The rate of change depends on the water vapor content and the momentary gain or loss of thermal energy at all altitudes. The change in temperature is called the lapse rate. The flow of energy through the universe constantly changes the lapse rate of a locality during the 24-h daily cycle.

Our weather systems are powered by nuclear-generated radiant energy from the sun. The atmosphere directly absorbs about 15% of the energy received from the sun, 42% reflects back into outer space and 43% is absorbed by the earth's surface. The earth converts the energy from short wave light energy to long wave heat energy. The long wave energy warms the troposphere and makes it move. The heat energy is ultimately lost into outer space. The earth acts like a greenhouse, passing short wave radiant energy and absorbing long wave heat energy.

The earth spins on its axis and turns around the sun, giving rise to the turbulence that results in wind patterns. Because the earth turns on its polar axis and its linear speed varies from about 7800 km an hour at the equator to zero at the poles, this motion changes the hot spots on the earth and moves the troposphere in six major zones, each with its prevailing wind motion or direction. The height of the troposphere varies from about 83 km in the tropics to about 38 km at the poles. The width of the zones varies with the season and the day. Air is a mixture of gases and it can therefore move independently of the earth. If the air moves with the earth we experience a calm, if it moves in any other direction, wind is generated. To further complicate the picture, great air masses move independently of one another; polar continental and maritime masses from the ends of the earth, tropical continental and maritime masses from the torrid tropical zone. The solar energy received by the earth waxes and wanes, causing changes in the dominance of the temperate latitudes by the hot and cold air masses. This we perceive as alternately fair or foul weather, with occasional periods of stagnation which we describe as drought or long periods of rainfall.

While all weather depends upon these world-wide forces, local topography can greatly alter or affect the local climate. Therefore any discussion of air pollution must consider the topography and the proximity to bodies of water which can influence local climates.

The foregoing discussion illustrates, in a very general way, how the weather patterns are formed and allows one to perceive how pollutants emitted into the troposphere might be dispersed and transported across the earth's surface. It is now appropriate to try to answer the two questions posed earlier in the chapter, namely—(1) which air pollutants affect plant ecosystems and (2) where do they come from?

II. WHICH AIR POLLUTANTS AFFECT PLANT ECOSYSTEMS?

What we now call 'air' consists of 21% oxygen, 78% nitrogen, about 0·03% carbon dioxide, less than 1% argon and traces of other gases, plus varying amounts of water vapor. This is the mix in which the living organisms of the earth can best function. For millions of years the air has been contaminated by natural products including salt droplets from the sea, ice, soil, dust, plant and animal propagules and parts, oxides of nitrogen and sulfur, methane, ammonia, hydrogen chloride and hydrogen fluoride, to name just a few. Soil erosion, volcanic eruptions, the wave motion of the ocean and billions of respiring, photosynthesizing, dying and decomposing organisms still fill the air with these same contaminants. So why the concern? What is new?

Man's technology is what is new. The iron plow, the massive burning of fossil fuels to heat space and water, drive machinery, make electricity; the manufacture of thousands of synthetic compounds hitherto unknown in nature and the unchecked rise in the numbers of humans on earth are the

TABLE 1

MAJOR AND MINOR AIR POLLUTANTS AND POLLUTANT MIXTURES THAT AFFECT PLANT ECOSYSTEMS

Pollutant	*Common source*	*Phytotoxic conc. (ppm/h)*
I. Major pollutants		
A. Photochemical oxidants		
1. Ozone	Internal combustion engine	0·04–0·7
2. Nitrates	Internal combustion engine	0·004–0·01
3. Oxides of nitrogen	Combustion furnace and internal combustion engine	0·21–100
B. Sulfur oxides	Combustion furnace (Coal or oil) Other commercial processes	0·1–0·5
C. Fluorides	Production of aluminum, phosphates, brick-making, sintering	0·0001
II. Minor pollutants		
1. Ammonia	Various industrial processes–spills	
2. Boron	Same	
3. Chlorine	Same	
4. Ethylene and propylene	Internal combustion engine	0·0005–10
5. Hydrogen chloride and hydrochloric acid	Various industrial processes	
6. Particulates and heavy metals	Combustion furnace Various industrial processes	Varies
7. Sodium sulfate	Combustion furnace	
III. Pollutant combinations	All of above sources	
1. Ozone and sulfur dioxide		
2. Ozone and peroxyacetyl nitrate		
3. Sulfur dioxide and nitrogen dioxide		

difference. To feed, clothe, house and transport these human hordes has resulted in the global exploitation of the earth's resources and the perceived waste products of this enterprise are disposed of in the oceans of air and water.

Specifically, the pollutants discussed here are either gaseous or particulate. They are further divided into major and minor pollutants and pollutant mixtures, as shown in Table 1. Not discussed here are pesticides, saline aerosols, acid rain (acidic precipitation) and a myriad of synthetic chemicals that do not appear to affect vegetation when present at their usual ambient concentrations.

The sources of air pollutants may be listed as: (a) stationary combustion sources, (b) mobile combustion sources, (c) petroleum refinery emissions, (d) the non-metallic mineral products industry, (e) ferrous metallurgical processes, (f) non-ferrous metallurgical processes, (g) inorganic chemical industries, (h) pulp and paper industries and (i) the food and feed industry.

A. Stationary Combustion Sources

Stationary combustion sources include fossil fuel burning in home and industry space heating and in electric power production. Pollution from this type of source results from unburned fuel particles and gases escaping from the furnace into the atmosphere. The following is a simplified description of furnace combustion (stationary) using coal or oil, both sulfur-containing fossil fuels. The prime reaction is the oxidation of carbon (carbon (C) + oxygen (O_2) = carbon dioxide (CO_2)). Actually, though, there are usually two additional reactions: $C + O_2 = CO_2$ and $2\,CO + O_2 = 2\,CO_2$. The rate of these reactions depends upon the temperature and degree of mixing. With adequate mixing, which is seldom achieved, below 1204 °C the outer layer of the carbon particle absorbs oxygen until saturated. Then outer molecules break off and combine with more oxygen to form equal amounts of CO and CO_2. The solubility of oxygen in carbon declines rapidly above 1204 °C and the reaction rate approaches zero at 1316 °C. Between 1316 °C and 1482 °C there is *no* active oxidation. This 166 °C span is simply a dead zone through which the reaction temperature rises at the start of combustion and drops through again at the end of the burning process.

Above 1482 °C carbon and oxygen combine directly without prior absorption. In this temperature range oxidation produces about twice as much CO and CO_2 because of the equilibrium coefficients for the extra reaction. Thus, the combustion sequence is as follows. (1) Oxidation to equal parts of CO and CO_2 up to 1316° C. (2) Very slow oxidation of the remaining carbon up to 1482 °C with conversion of CO to CO_2 dependent upon the hydrogen (H_2) and H_2O content of the gases at that time. The presence of H_2 speeds the reaction of CO to CO_2. (3) Carbon oxidation continues after the temperature drops below 1316 °C. The burning of carbon and the conversion to CO_2 slows down as the reaction temperature drops below 1204 °C. The result is that unburned carbon and a considerable amount of CO not converted to CO_2 show up as particulate and gaseous pollutants in the emission gases escaping from the stack.

Even total combustion can create pollution problems. Good combustion usually results in small amounts of unburned carbon and traces of CO coupled with sulfur compounds (SO_x) and oxides of nitrogen. Sulfur present in the fuel converts to about 98% sulfur dioxide (SO_2) and 2% sulfite (SO_3) during normal combustion. In gaseous form these combine with water vapor in the flue gas or outside atmosphere to produce acids. Coal firing results in more sulfur remaining in the ash. Nitrogen oxides (NO_x) also form under excellent combustion conditions. Nitrogen oxide (NO) can exist at high temperatures and concentrations increase as gas temperatures rise. The per cent NO increases along with excess air. NO is the primary oxide of nitrogen since the high furnace temperatures promote nitrogen fixation in combustion air according to the equation: $N_2 + O_2 = 2NO$. At lower temperatures further oxidation produces nitrogen dioxide (NO_2), one of the primary precursors of photochemical oxidant smog. Flue gas is hot enough to hold NO_2 to a minimum in the stack. However, once the gas leaves the stack, lower atmospheric temperatures and an abundance of oxygen promote NO_2 production. An exit temperature of 1093 °C, followed by rapid cooling in the atmosphere produces 1000 ppm NO which rapidly converts to large amounts of NO_2. Figures generated by the TVA (Tennessee Valley Authority) reveal that, during an average year in the area under TVA jurisdiction, about 4500 tons of NO_2 are emitted from TVA combustion stacks. Many of these stacks are over 200 m tall which ensures that these pollutants are able to enter the larger air masses, which travel long distances on the prevailing winds, thus ensuring that the pollutant or pollutant products will impact vegetation at very long distances from their source.

In summary, stationary combustion sources produce CO, CO_2, SO_x, NO_x and particulates of carbon fly ash which may also contain some trace metals and non-combusted organic molecules adsorbed onto their surfaces.

B. Mobile Combustion Sources

With the disappearance of the steam locomotive, mobile combustion sources comprise all of the vehicles that utilize some form of the internal combustion engine which may be driven by some type of hydrocarbon liquid fossil fuel-derived product such as gasoline or diesel fuel. These are fuels based on crude oil derivatives. The pollutants generated by these engines are also the product of carbon combustion, but are unique because of the complex collection of organic hydrocarbons which they contain.

The internal combustion engine is much less efficient than a large stationary combustion furnace because: (a) the fuel/air ratios are more variable, (b) acceleration and deceleration modify this ratio and (c) the running temperatures are always lower. The chief constituents of car exhaust gases are CO, CO_2 and water vapor. Lower concentrations of partially burned or unburned hydrocarbons and nitrogen oxides are also produced. NO concentrations vary between 1000 and 3000 ppm as the engine is speeded up. The highest hydrocarbon concentrations occur in the exhaust when the engine is slowed, resulting in levels as high as 4000–12000 ppm.

Both stationary (combustion furnace) and mobile (internal combustion engine) sources give rise to the NO_x series but mobile sources generate almost all of the hydrocarbons. This means that both sources can contribute to the formation of the photochemical pollutants which are the major ingredients of what is now known as 'smog'.

The photochemical oxidants encountered in the atmosphere are ozone (O_3), nitrogen oxides (NO_x) and peroxyacyl nitrates such as PAN. These compounds form a complex series as shown in Table 2, Chapter 2.

Oxidants are secondary pollutants since they are formed as a result of chemical reactions in the atmosphere. Primary pollutants, as previously noted, are those that are emitted directly by the pollution source. The primary pollutants responsible for the formation of photochemical oxidants in air are nitric oxide (NO), hydrocarbons (CH_n), aldehydes (CHO), carbon monoxide (CO) and carbon dioxide (CO_2). Both stationary and mobile sources produce some of these pollutants, but mobile sources produce all of them. The photochemical oxidants of greatest significance to plant growth are ozone, nitrogen dioxide and peroxyacetyl nitrate (PAN). The major hydrocarbon pollutant that affects plants directly (primary) is ethylene (C_2H_4).

The amount of oxidant found in the atmosphere depends upon the time of day, meteorologic conditions and the available amounts of primary pollutants. Early studies showed a diurnal pattern with low oxidant levels in the morning, rising to a peak about 2–3 pm, and then dropping. However, newer monitoring studies indicate that oxidant production may peak later in the day and often continues well into the night before declining to baseline levels.

The major oxidant in photochemical smog is ozone and this is also the most phytotoxic material in the smog mix. Its formation is still not clearly understood after at least 25 years of study. Twenty-five equations were developed in the National Academy of Sciences review, which then summarized these by stating: 'the concentration of ozone in the polluted atmosphere is controlled by the intensity of sunlight and the ratio of nitrogen dioxide to nitric oxide (NO_2/NO)'.

Hydrocarbons and other pollutants such as aldehydes, ketones, chlorinated hydrocarbons and carbon monoxide react to form peroxy radicals. These, in turn, react with nitric oxide, causing the ratio NO_2/NO to increase. This allows the ozone concentration to increase as the consequence of a change in the equation $O_3 = (0{\cdot}021\ \text{ppm})\frac{NO_2}{NO}$. Basic to the functioning of this dynamic system are the reactions: (1) $NO_2 + \text{light}$ ($\lambda < 430$ nm) $= O + NO$ and (2) $O + O_2 + M = O_3 + M$, where M can be a nitrogen or oxygen molecule. Put the two reactions together and add a third; (3) $NO + O_3 = NO_2 + O_2$, and you have the basis for ozone formation in the atmosphere. The further addition of hydrocarbons, aldehydes and hydroxyl (OH) and CO into the mix ensures, through another series of reactions, that more NO_2 will be produced without going through the third, thus allowing a build up of O_3 which proceeds faster than the destruction of O_3 through the third equation. This permits the ozone level to increase beyond the levels predicted by the first series of equations (1) to (3).

Peroxyacetyl nitrate (PAN) is also formed by a series of chemical reactions similar to those producing ozone, but the concentration of ambient ozone is usually about 10–100 times greater than that of PAN.

Many of the hydrocarbons emitted by the internal combustion engine are significant because they play a role in the formation of photochemical oxidants or because they participate in the formation of aerosols which may be eye or respiratory irritants. By themselves, they tend not to be highly phytotoxic. Ethylene, an exhaust hydrocarbon, does affect vegetation at concentrations of a few parts per billion (ppb)†. It is a primary pollutant and is therefore considered in chapter 8. It may make up as much as 20–30% of the total hydrocarbon exhaust emissions.

† Throughout this book 'billion' represents 10^9.

Lead is a primary particulate pollutant emitted by gasoline-burning engines. It is added to gasoline to increase the power output of the burned fuel and is distributed alongside roadways as larger particles and can also become airborne as a lead aerosol which may then be more widely dispersed away from the source.

C. Other Pollutant Sources

Pollution from sources other than the combustion of fossil fuels include:

(1) Pulp and paper manufacturing.
(2) Food and feed industries.
(3) The petroleum refining industry.
(4) The inorganic chemical industry.
(5) Ferrous metallurgical industries.
(6) Non-ferrous metallurgical industries.

All of these activities contribute a spectrum of gaseous and particulate pollutants. The character and quantity of the pollutant mix depends mostly on the raw materials entering the process and the form of the process. Pollution from these sources tends to be fairly localized, but may travel some distance on favorable wind patterns.

Pollutants contributed by these sources include nitrogen oxides, sulfur and carbon, as well as sulfur dioxide and carbon monoxide and dioxide; a variety of fluorides, including both gaseous and particulate forms; trace metals such as zinc, copper, nickel, cadmium, silver, arsenic, aluminum, iron and vanadium, just to name a few. A variety of inorganic acids are also produced, but many of these are recycled. Finally, large amounts of coarse and fine dust particulates are produced, some of which are contaminated by organic materials or trace metals.

Because of the nature of many of these processes, effective control can be exercised at the source, reducing the amount of pollutants that can escape into the atmosphere. This fact holds true for many of the so called 'minor pollutants' like hydrochloric acid, chlorine gas, ammonia gas, boron, etc. which tend to pollute only as the result of accidental spillage, either during loading or transporting the materials. The effects may be quite severe but tend to be very site-specific unless the spill happens to allow for contamination of a waterway.

The sources responsible for mixes of pollutants are the same sources already discussed in this chapter. The presence or absence of a particular pollutant in the air is related to the presence or absence of a particular pollutant source or group of sources, as well as the location of the impacted site with relation to the prevailing winds drawing pollutants from distant sources. The air is always loaded with a pollutant mix. The three mixes we have selected for treatment here are ones which have been studied enough to be able to predict with some certainty how plants will react in their presence.

III. SUMMARY

Air is a mixture of gases and naturally produced pollutants which coexist with man-made pollutants generated by the stationary and mobile combustion of fossil fuels, plus the effluents from a variety of processing and manufacturing industries. Pollutant concentration and movement are dependent upon local and global meteorological patterns, energy derived from the sun and a complex mix of combustion and air chemistry. We describe primary and secondary pollutants, as well as gaseous and particulate pollutants. Rural and urban areas are both impacted by pollutants because of dispersion patterns and the dynamics of pollutant chemistry.

BIBLIOGRAPHY

The following are suggested for readers who wish to pursue further the introductory material presented in this chapter.

1. FRIEDLANDER, S. K. (1977). *Ozone and Other Photochemical Oxidants.* Committee on Medical and Biologic Effects of Environmental Pollutants of the National Academy of Sciences. Washington, DC, 719 pp.
2. HECK, W. W., S. V. KRUPA and S. N. LINZON. (1978). *Handbook of Methodology for the Assessment of Air Pollution Effects on Vegetation.* Air Poll. Contr. Assoc., Pittsburgh, Pa. 392 pp.
3. LACASSE, N. L. and W. J. MOROZ. (1969). *Handbook of Effects Assessment*, Vegetation Damage Center for Air Environment Studies. Penn. State Univ., Univ. Park, Pa. 193 pp.
4. MUDD, J. B. and T. T. KOZLOWSKI (eds.). (1975). *Responses of Plants to Air Pollution.* Academic Press, New York, 383 pp.
5. STERN, A. C. (1976). *Air Pollution.* (3rd. ed.), Vols. I–V. Academic Press, New York, 3844 pp.
6. Symposium Proceedings (1968). Trends in air pollution damage to plants. *Phytopathology*, **58**: 1075–113.
7. Symposium Proceedings (1969). Pollutant impact on horticulture and man. *HortScience*, **5**: 237–52.

A common symptom of O_3 injury on plants is called 'flecking'. It is usually an indication of acute injury. Small necrotic spots or flecks occur due to death of palisade cells. These flecks are metallic or brown and often bleach to tan or white with age. Flecks may coalesce to form blotches which can result in chlorosis and leaf fall (Figs 1 to 3).

Stipples are a restricted form of acute O_3 injury. Stipples are punctate spots consisting of a few palisade cells that have been injured or killed by O_3. They may be white, black, red or reddish–purple (Figs 4 and 5).

Many leaves take on an extensive reddish–brown or bronze color when exposed to low concentrations of O_3. This usually leads to chlorosis, senescence and leaf fall (Fig. 6).

Chlorosis of older leaves may be the only symptom of chronic O_3 injury (Fig. 7). This occurs over long time periods and results in premature defoliation.

On conifers, needle tips may turn red, then brown, and fade to grey. Isolated yellow spots or mottling may occur in the needles.

FIG. 1. Weather-fleck on older leaves of Bel-W3 tobacco (*Nicotiana tabacum* L.), caused by ozone (O_3). (Courtesy: USDA.)

FIG. 2. White, bleached flecks on older leaves of spinach (*Spinacia oleraceae* L.), caused by ozone (O_3). (Courtesy: USDA.)

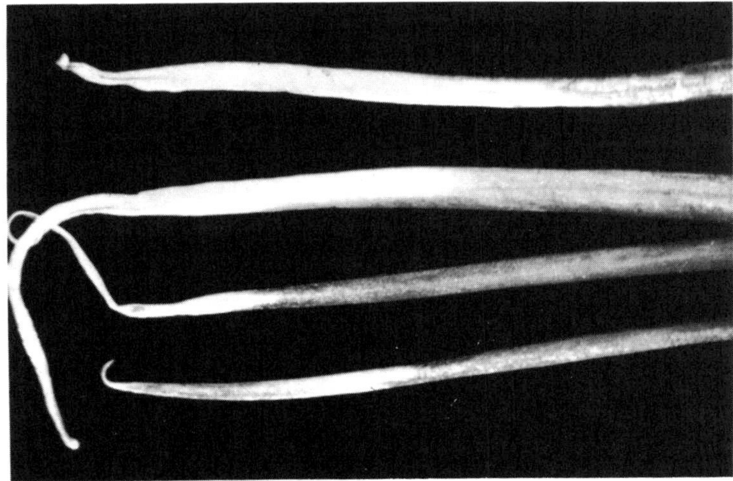

FIG. 3. White, bleached tipburn on older leaves of onion (*Allium cepa* L.), caused by ozone (O_3). (Courtesy: US EPA.)

FIG. 4. Reddish-purple, punctate stipple on older leaves of ash (*Fraxinus americana* L.), caused by ozone (O_3). (Courtesy: US EPA.)

FIG. 5. White, punctate stipple on older leaves of squash (*Cucurbita maxima* Dcne.), caused by ozone (O_3). (Courtesy: USDA.)

FIG. 6. Bronzing and chlorosis on older primary leaf of Pinto bean (*Phaseolus vulgaris* L.), caused by ozone (O_3). (Photo: William J. Manning.)

FIG. 7. Chlorosis on older leaves of poinsettia (*Euphorbia pulcherrima* Willd.), caused by ozone (O_3). (Photo: William J. Manning.)

TABLE 3

PLANTS COMMONLY AFFECTED BY OZONE AND TYPICAL SYMPTOMS EXPRESSED

Plant[a]	*Typical symptoms*
Ash (*Fraxinus*)	White stipple, purple bronzing
Bean (*Phaseolus*)	Bronzing, chlorosis
Cucumber (*Cucumis*)	White stipple
Grape (*Vitis*)	Red–brown to black stipple
Morning glory (*Ipomoea*)	Brown fleck, chlorosis
Onion (*Allium*)	White fleck, bleached tips
Pine (*Pinus*)	Tipburn, needle mottle
Potato (*Solanum*)	Grey, metallic fleck
Spinach (*Spinacia*)	Grey–white fleck
Tobacco (*Nicotiana*)	White–grey fleck
Watermelon (*Citrullus*)	Grey, metallic fleck

[a] Not all varieties of these plants are O_3-sensitive. Specific ones must be grown to obtain these responses.

Some plants commonly affected by O_3 are listed in Table 3. Typical symptoms are given for each plant. Depending on the variety used, weather conditions, O_3 concentration and duration, leaf age, etc., it would be possible to find varying degrees of both acute and chronic O_3 injury on these plants in the field.

2. *Peroxyacyl nitrates*

A number of peroxyacyl nitrates occur in the ambient oxidant complex (Table 2). Of these, peroxyacetyl nitrate, or PAN, is the best known and appears to be the most phytotoxic. It is formed as a secondary product from a complex reaction between hydrocarbons and sunlight.

PAN enters the stomates of the youngest, most actively growing, leaves on plants that are also young and actively growing. Injury often occurs at the apex of the youngest leaves and the base of older, yet sensitive, leaves.[35] Injury occurs most frequently when plants are exposed to PAN at high temperatures rather than low ones.

Injury symptoms vary with plant type and may be confused with those caused by other pollutants, pollutant mixtures, frost, or other factors.[18,19,30,35,45,48,55–57]

On plants with broad leaf blades, water-soaked spots may occur on the undersurface of leaves. This later may be glazed or silvery or may become bronzed (Fig. 8). Romaine lettuce, Pinto bean and Swiss chard in

FIG. 8. Undersurface glazing and bronzing on primary leaf of Pinto bean (*Phaseolus vulgaris* L.), caused by peroxyacetyl nitrate (PAN). (Courtesy: US EPA.)

California demonstrate these symptoms. Certain petunias, however, may show apical or basal leaf bleaching that is bifacial and could be confused with O_3 or SO_2 injury. On narrow-leaved plants, such as grasses and grains, chlorotic or bleached bands may appear in the leaves.[35]

PAN injury occurs frequently on vegetation on the west coast of the USA. PAN-like injury occurs on vegetation infrequently on the east coast. It is likely that PAN occurs on the east coast, but this has not been confirmed.

3. *Oxides of nitrogen*

There are a number of oxides of nitrogen in polluted air (Table 2). In addition to being involved in the reactions that lead to O_3 formation, two of them are thought to be potential phytotoxic air pollutants. Nitric oxide (NO) and nitrogen dioxide (NO_2) have been studied together as NO_x. NO_2 and NO have also been investigated as separate air pollutants. In general, higher concentrations of NO_x are required to cause plant injury than other pollutants, such as O_3 or SO_2. Many ambient concentrations of NO_x are too low to cause visible injury.

NO_x and NO_2 have been investigated the most.[3,16,54,57,60,65,69] There

FIG. 9. Marginal bleaching on leaves of periwinkle (*Vinca rosea* L.), caused by nitrogen dioxide (NO_2). (Courtesy: US EPA.)

are no reliable or typical diagnostic symptoms or common bioindicator plants. Acute NO_2 injury may resemble acute SO_2 injury (Fig. 9). At low concentrations, growth stimulation may occur and plants are a darker green in color. A non-specific chlorosis may occur which is followed by premature leaf fall. Reduction in young plant growth rate and dry matter accumulation is a common response in tomato.

Anderson and Mansfield[3] have recently found that NO is the principal air pollutant in glasshouses in the UK where hydrocarbons are burned for heat or CO_2 sources. Tomato varieties differed in their response to NO. The variety Sonato was stimulated by 0·40 ppm NO, especially if soil fertility was low. Growth inhibition occurred between 0·40 and 0·80 ppm NO.

B. Sulfur Dioxide

Sulfur dioxide (SO_2) is a well-known air pollutant that has been studied as a phytotoxicant for many years. Its association with power plants (especially those burning coal) and manufacturing operations usually allows a dispersal pattern that shows high concentrations near the source and a gradation of response away from it. A combination of acute and chronic SO_2 injury should be expected in the field.

A great deal is known about how SO_2 injures leaves. SO_2 enters via stomates, SO_2 is oxidized to sulfite (SO_3) and then slowly to sulfate (SO_4). SO_3 is highly toxic. SO_4 is less toxic. At low concentrations, SO_3 conversion occurs and injury is avoided. At higher concentration for longer durations, SO_2 conversion to SO_3 occurs faster than conversion of SO_3 to SO_4 and leaf injury results. SO_4 concentrations in plants can also increase to phytotoxic levels over prolonged exposure times.[35] Leaf analysis can be used to determine sulfur accumulation in plant tissues.

Acute SO_2 injury on broadleaf plants consists of interveinal bleaching (brown or white) and marginal bleaching (Fig. 10). On some leaves, a 'herring bone' effect may be observed, with the veins as the 'bones' (Figs 11 and 12). This bleaching is bifacial and begins as water-soaked dark areas on younger fully expanded leaves. Chronic or SO_4 injury may consist of chlorosis or red–brown discoloration. Coniferous needles exhibit reddening from the tips back to the base. SO_2 and SO_4 symptom expression has been extensively described and illustrated.[9,19,23,24,25,34,44,46,66]

SO_2 is known to cause plant injury at dosages of 0·05 to 0·50 ppm for 8 h

FIG. 10. Marginal bleaching of alfalfa leaves (*Medicago sativa* L.), caused by sulfur dioxide (SO_2). (Courtesy: USDA.)

FIG. 11. Interveinal bleaching on dewberry leaves (*Rubus* sp.), caused by sulfur dioxide (SO_2). (Courtesy: US Forest Service.)

FIG. 12. Interveinal browning on fox grape leaves (*Vitis vulpina* L.), caused by sulfur dioxide (SO_2). (Courtesy: US EPA.)

TABLE 4

INDICATORS OF SO_2 INJURY AND HOW THEY RESPOND IN THE FIELD

Plants[a]	*Responses*
Deciduous	
Brambles	
Blackberry	Interveinal, brown bleaching
Raspberry (*Rubus* spp.)	
Ferns	
Bracken (*Pteridium* spp.)	Reddish marginal necrosis
Hay-scented (*Dennstaedtia* spp.)	
Woody perennials	
Sweet Birch (*Betula lenta* L.)	Marginal and interveinal bleaching
White Ash (*Fraxinus americana* L.)	Extensive interveinal bleaching
Evergreens	
Pines	
Austrian (*Pinus nigra* Arnold)	Bands of necrosis on needles
Scotch (*Pinus sylvestris* L.)	Needle tips browned
White (*Pinus strobus* L.)	All of needle browned
Spruces	
Blue (*Picea pungens* Engelm.)	Needles brown and fall off
Norway (*Picea abies* L.)	

[a]Most are best early in the season. They become more resistant to SO_2 as the season progresses.

or more.[44] Table 4 lists some indicators of SO_2 injury and how they respond in the field.

C. Fluorides

Atmospheric fluoride is unique in that it can occur in three forms: as a gas, as particulate matter or as gaseous fluoride adsorbed to other particulate matter. Gaseous hydrogen fluoride (HF) is more toxic than particulate fluoride. As a result, more is known about HF than other fluoride forms.[35]

HF is a 'point-source' pollutant from smelting and other manufacturing operations, usually involving aluminum. Vegetation nearest the point of emission should show the most severe acute injury.

Chronic HF injury may result in chlorosis or chlorosis along leaf veins (Fig. 13). Acute HF injury results in marginal necrosis that begins at leaf tips and progresses to leaf bases. Leaves may be deformed due to apical and marginal necrosis (Figs 14 and 15).

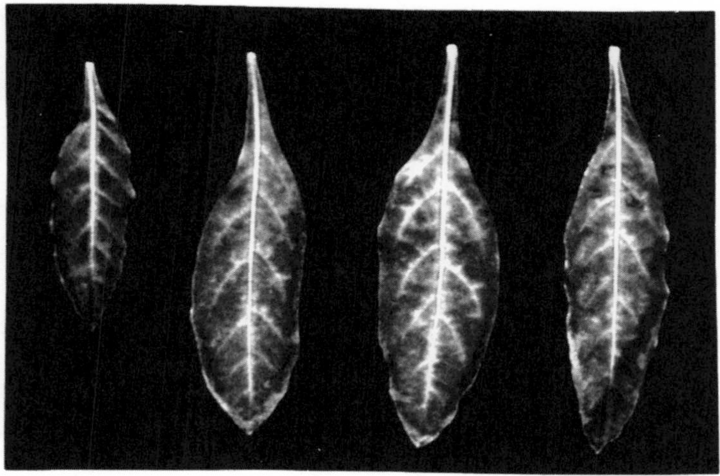

FIG. 13. Chlorosis along leaf veins of Jerusalem cherry (*Solanum pseudocapsicum* L.), caused by hydrogen fluoride (HF). (Courtesy: USDA.)

FIG. 14. Tip and marginal necrosis of Jerusalem cherry leaves (*Solanum pseudocapsicum* L.), caused by hydrogen fluoride (HF). (Courtesy: USDA.)

FIG. 15. Tip and marginal necrosis of birch leaves (*Betula* sp.), caused by hydrogen fluoride (HF). (Courtesy: USDA.)

HF symptoms on plants have been extensively investigated and described.[14,17,41,53,63,64,68] On broadleaved plants, necrosis begins at leaf tips and expands to leaf margins. Gladiolus is a popular monocot indicator plant. Depending on the variety used, white to tan to brown necrosis occurs from leaf tips downwards. A dark brown, distinct band separates dead and healthy tissue. Tulip, iris and lily show similar symptoms. Conifer needles exhibit burned tips or marginal to complete scorch.[35]

HF is unique in that it accumulates in leaves, particularly at the margins and apices. Leaf tissue analysis is commonly used to assess HF effects on vegetation.

III. MINOR POLLUTANTS

A. Ammonia

Occasional industrial accidents, pipeline disruption or carrier accidents result in the release of ammonia (NH_3) into the atmosphere. Plants near the spill may suffer severe acute injury. Like NO_x, high NH_3 concentrations are required to injure plants.

Little work has been done with NH_3 and plants. Middle-aged leaves appear to be the most sensitive and the entire leaf may turn dull green and then brown or black. Leaf pH may increase and account for the color change. Lower concentrations of NH_3 may cause glazing or silvering on the undersides of leaves, which could be confused with PAN injury.[35] It has also been noted that NH_3 can cause purple to black areas around lenticels on apples.[29]

B. Boron

Temple and Linzon[58] investigated boron emissions from factories manufacturing fiberglass, stoves and refrigerators in Ontario, Canada. On some plants near the sources, obvious leaf cupping and distortion occurred, as did marginal and interveinal necrosis and flecking. Injury was most severe on the older leaves. These effects were most severe within 200 m of the sources and decreased significantly at 500 m. Buckthorn, butternut, hackberry, honeysuckle, maple, mulberry, spiraea and wild grape were susceptible while elm, lilac, pear, privet, tree-of-heaven and most herbaceous plants showed little, if any, injury.

C. Chlorine

Chlorine (Cl_2) is a widely used oxidizing agent. It is commonly transported by truck or rail car and accidental spills can cause severe acute injury in the region of the spill.

FIG. 16. White stipple on horsechestnut leaves (*Aesculus hippocastanum* L.), caused by chlorine (Cl_2). (Courtesy: USDA.)

Symptoms of Cl_2 injury have been described on a number of plants.[4,11,12,35] Very dark green to black spots occur near margins which later fade to white or brown. Interveinal injury of this type may resemble that caused by SO_2. A fleck–stipple effect may also occur, which resembles O_3 injury (Fig. 16). Tip necrosis and mottling can occur on conifers, which resembles O_3 injury.[35] Mustard, chickweed and sunflower are sensitive to Cl_2.

D. Ethylene and Propylene

Ethylene (C_2H_4) and related olefins, such as acetylene and propylene, are a group of gases that are becoming more important as phytotoxic air pollutants.

Ethylene is a natural plant hormone that forms when plants are injured (even by other air pollutants). It plays an important role in flowering, fruit ripening, senescence and abscission.[1] It is also becoming a contaminant in the atmosphere, with motor vehicle exhaust as its primary source.[2]

C_2H_4 actively affects plants at concentrations expressed in parts per billion (ppb). Most other gaseous air pollutants are expressed in parts per hundred million (pphm) or parts per million (ppm). Symptoms include epinasty in young leaves of plants such as tomato, reduced growth, premature senescence and abscission, reduced flowering and fruit set, premature bud break, inhibition of leaf expansion and leaf curling. In orchids, acute injury occurs on the sepals. They turn dry and brown and may bleach.[35] Carnation flowers begin to close and are said to be 'sleepy' or 'bullheads'.[23] A threshold concentration for adverse growth effects of C_2H_4 on some plants is 10 ppb.

Propylene is an olefin related to ethylene. It has been found in glasshouses in England as a contaminant of liquefied propane gas used to supply CO_2. Propylene effects appear to resemble those caused by C_2H_4, except that higher concentrations are required. Propylene inhibits flower production in chrysanthemum (variety 'Polaris') and encourages leaf formation. Apical dominance is decreased and many side buds develop into short leafy shoots. Propylene-treated plants had shorter internodes and smaller, thicker leaves than non-treated plants. This work was done by Dr David Hand at the Glasshouse Crops Research Institute.

E. Hydrogen Chloride and Hydrochloric Acid

Hydrogen chloride is occasionally released in the atmosphere at local point sources. It is a very hygroscopic gas and could change to hydrochloric acid aerosol droplets in the atmosphere.[5,35]

FIG. 17. Red–brown to black spotting and 'shot-holing' on leaves of black cherry (*Prunus serotina* Ehrh.), caused by hydrogen chloride (HCl). (Courtesy: US EPA.)

A wide variety of symptoms have been reported. Marginal interveinal chlorosis, followed by yellow, brown, red to black necrosis is a common response. White to cream-colored margins may occur around necrotic areas. Pan-like injury can occur on tomato leaves. Flecking or spotting of upper leaf surfaces, in obvious red–brown to black colors, may be symptoms of hydrochloric acid aerosol injury. Shot-holing, or loss of the tissue in the centers of the spots, suggests an acid injury effect. Symptoms like this have been observed on black cherry leaves (Fig. 17).

Chlorides often accumulate in leaf tips, like fluorides. Leaf analysis would allow detection of chloride levels in injured leaves.

F. Particulates and Heavy Metals

Air contains many particles which are continuously deposited on plant surfaces. Some are blown or washed off, others enter leaves through stomates or by injuring epidermal cells.

These small, multi-molecular units are measured in microns and sorted out by particle size. Determining size can be difficult as they interact with each other, water and gases. In themselves, they often tend to be rather inert, but, when combined with other substances, they may be phytotoxic. HF and SO_2 are water-soluble gases that can become part of water films around particles. SO_2 dissolution can result in acidic particles which burn plant leaves.[35]

A number of investigators have reported the effects of airborne particulates on plants under natural conditions.[6,13,23,24,25,31,36,40,43,52,69] Particulates may form physical barriers on leaves that decrease light availability and hence photosynthesis.[36] Particulates may clog stomates and increase SO_2 susceptibility[52] or interfere with flower pollination,[36] leaf size and condition[43] and the composition of forest stands[13] by affecting soil pH. The role of particulates in air pollution effects on vegetation needs a good deal more investigation.

Most heavy metals occur in the atmosphere in some type of particulate form as salts or adsorbed to other particles. Deposition occurs on plant parts and the soil surface. There is considerable debate as to whether heavy metals are absorbed by leaves in any appreciable way or whether they are absorbed by roots and stored or translocated upward to leaves, fruits, etc.

Heavy metals that fall on soil tend to accumulate in upper soil layers. Clay and organic matter content will affect the availability of heavy metals. In general, heavy metals are stable and not leached or degraded. When concentrations rise due to continuous application, toxic concentrations may occur. Selection for resistant plants also occurs.[69]

Lead (Pb) is the most common and most abundant heavy metal in air and soil. Most Pb emissions come from leaded gasoline combustion and manufacturing operations.[35] Paints used to contain Pb.

There is considerable controversy with regard to whether or not Pb in plants comes from foliar entry, root entry, or both, whether Pb is translocated in plants and whether it adversely affects plants.[10,15,20,42,62,71] Pb does occur on leaves, but most of it can be removed by washing. Pb is taken up by plant roots, but is thought to be localized in dictyosome vesicles and deposited in cell walls. Pb does accumulate in soil, but no clear evidence of Pb poisoning, in plants grown under natural conditions, has been reported. This is an area that needs considerably greater investigation.

Zinc, cadmium and copper were described as causing foliar interveinal chlorosis, followed by reddening and yellowing, on leaves of trees near the source in mid-summer.[35]

Mercury (Hg) is a unique heavy metal in that it is a liquid at normal temperatures. In a closed greenhouse environment, toxic vapors from Hg-containing paint can adversely affect many plants, especially roses. Brown spots occur on rose leaves, leaves yellow and then abscisse. Young buds brown and fall. Flower petals fade and turn brown and stamens may be killed.[35]

The detection of heavy metals in soil and plant tissues is quite possible, using techniques such as atomic absorption spectrophotometry.

G. Sodium Sulfate

Temple and Richards[59] found sodium sulfate (Na_2SO_4) in the atmosphere around kraft pulp and paper mills in Ontario, Canada. In greenhouse tests, they found growth stunting and foliar necrosis in Pinto bean and height reduction in 'Veemore' tomato.

IV. POLLUTANT COMBINATIONS

The atmosphere around plants in the field usually contains several potential phytotoxic air pollutants. How these pollutants interact and how this affects the expression of injury on plants has not been extensively explored. It has long been suspected, however, that some air pollutant symptoms are really due to a mixture of gases rather than just one. Mixtures of gases may also result in symptoms that resemble those caused by single pollutants. Mixtures of gases may also affect the threshold level where plants become sensitive to one or both of the pollutants. Two gases mixed together may cause more or less injury than either of the two alone (synergism).

Almost all work on pollutant combinations has been done under controlled experimental conditions.[50] Some examples are given below:

A. Ozone and Sulfur Dioxide

Jacobson and Colavito[32] described tan to white interveinal necrotic lesions on bean and tobacco plants exposed to O_3 and SO_2 (Fig. 18). In mixtures used, symptoms were similar to those caused by O_3 or SO_2, depending upon which pollutant was at a concentration above the threshold for injury. Reinert *et al.*[50] summarized several reports and concluded that if O_3 and SO_2 mixtures are below the threshold for SO_2 injury, but at or below the threshold for O_3, foliar symptoms observed will be similar to those caused by O_3. Plants for which greater than additive injury effects for O_3–SO_2 mixtures occurred included: alfalfa, broccoli, cabbage, onion, Pinto bean, radish, spinach, soybean, tobacco, tomato and white pine. Lewis and Brennan[37] found that the foliage of petunia plants exposed to a mixture of O_3 and SO_2 developed symptoms of 'classic' PAN injury. They questioned whether PAN injury on petunia was really due to an O_3–SO_2 mixture.

FIG. 18. Interveinal bleached spots on tobacco leaves (*Nicotiana tabacum* L.), caused by a combination of ozone (O_3) and sulfur dioxide (SO_2). (Courtesy: US EPA.)

B. Ozone and Peroxyacetyl Nitrate

Davis[18] combined PAN and O_3 concentrations high enough to cause acute injury on young needles of Ponderosa pine. Similar seedlings were exposed to PAN and O_3 alone. Exposure to PAN + O_3 resulted in less injury than exposure to O_3 alone. PAN alone did not cause injury. This antagonistic interaction resulted in injury reduction.

C. Sulfur Dioxide and Nitrogen Dioxide

Tingey *et al.*[61] found that when SO_2 and NO_2 were combined at concentrations below the injury thresholds for each, upper surface injury occurred on oat, Pinto bean, radish, soybean, tobacco and tomato. On lower leaf surfaces, silvering or reddish pigmentation occurred. Ashenden and Mansfield[8] found that $SO_2 + NO_2$ combinations resulted in dry weight reductions in four pasture grasses, where yield reductions from either pollutant alone might not be expected.

V. DIAGNOSING AIR POLLUTION INJURY

A thorough knowledge of air pollutants and their usual or expected effects on plants is helpful, but it is not enough background to determine whether or not a plant has been injured by an air pollutant. Background and training are needed in agronomy, ecology, entomology, horticulture, plant nutrition, plant pathology and plant physiology as well. Time and experience in the field are also essential before any diagnoses can be made. A glance at Table 5 shows some of the agents that can cause symptoms that

TABLE 5

AGENTS THAT CAN CAUSE SYMPTOMS THAT RESEMBLE THOSE CAUSED BY AIR POLLUTANTS[a]

Living agents	*Non-living agents*
Bacteria	Light
Fungi	Mechanical injury
Insects	Nutrients
Mites	pH
Mycoplasma-like organisms	Pesticides
Nematodes	Salt
Viruses	Temperature
Genetic aberrations	Water
	Wind

[a]Adapted from: Lacasse, N. L. and M. Treshow, (Eds.). *Diagnosing Vegetation Injury Caused by Air Pollution*, US Environmental Protection Agency Handbook.[35]

TABLE 6

INFORMATION TO OBTAIN WHEN DIAGNOSING PLANTS IN THE FIELD

Crop history	*Plants observed*	*Geographical*
Pesticide usage	How many species affected	Topography of the area
Nutrient usage	Variety names (cultivated plants)	Wind patterns
Disease agents present	Parts of plants affected	Possible pollutant sources
Unusual weather	Locations of affected plants in relation to possible pollutant sources	
Cultural practices employed		

resemble those caused by air pollutants. This Table illustrates why one must be a good plant scientist to be a good air pollution diagnostician in the field.

A good case history must be obtained when diagnosing suspected air pollutant injury on native or cultivated plants in the field. The major components required to write a good case history are given in Table 6.

REFERENCES

1. Abeles, F. B. (1973). *Ethylene in Plant Biology*, Academic Press, NY.
2. Abeles, F. B. and H. E. Heggestad. (1973). Ethylene: An urban air pollutant. *J. Air Poll. Contr. Assoc.*, **23**: 517–21.
3. Anderson, L. S. and T. A. Mansfield. (1979). The effects of nitric oxide pollution on the growth of tomato. *Environ. Pollut.*, **19**: 113–21.
4. Anon. (1971). *Chlorine and air pollution: An annotated bibliography* US EPA Office of Air Programs, Pub. No. AP-99, Research Triangle Park, North Carolina.
5. Anon. (1971). *Hydrochloric acid and air pollution: An annotated bibliography.* US EPA Office of Air Programs, Pub. No. AP-100.
6. Anon. (1977). *Particulates and air pollution: An annotated bibliography.* US EPA Office of Air Programs, Pub. EPA-450/1–77–001.
7. Anon. (1979). Propylene pollution. *The Grower (London)* (22 Feb.), 38.
8. Ashenden, T. W. and T. A. Mansfield. (1978). Extreme pollution sensitivity of grasses when SO_2 and NO_2 are present in the atmosphere together. *Nature (London)*, **273**: 142–143.
9. Barrett, T. W. and H. M. Benedict. (1970). Sulfur dioxide. In: *Recognition of Air Pollution Injury to Vegetation: A Pictorial Atlas.* (Jacobson, J. S. and A. C. Hills (eds.)), Air Pollution Control Association, Pittsburgh, Pa., pp. Cl–C17.
10. Boggess, W. R. (ed.). (1977). *Lead in the environment.* US National Science Foundation, Pub. NSF/RA-770214.
11. Brennan, E., I. A. Leone and R. H. Daines. (1965). Chlorine as a phytotoxic air pollutant. *Int. J. Air and Water Poll.*, **9**: 791–7.
12. Brennan, E. and I. A. Leone. (1966). Response of pine trees to chlorine in the atmosphere. *Forest Science*, **12**: 386–90.
13. Brandt, C. J., and R. W. Rhoades. (1972). Effects of limestone dust accumulation on composition of a forest community. *Environ. Pollut.*, **3**: 217–25.
14. Brewer, R. F., F. B. Guillemet and F. H. Sutherland. (1966). The effects of atmospheric fluoride on gladiolus growth, flowering, and corm production. *Proc. Amer. Soc. Hort. Sci.*, **88**: 631–4.
15. Buchauer, M. J. (1973). Contamination of soil and vegetation near a zinc smelter by zinc, cadmium, copper and lead. *Environ. Sci. and Tech.*, **7**: 131–5.
16. Capron, T. M. and T. A. Mansfield. (1977). Inhibition of growth in tomato by air polluted with nitrogen oxides. *J. Expt. Bot.*, **28**: 112–6.
17. Chang, C. W. (1975). Fluorides. In: *Responses of Plants to Air Pollution.*, Academic Press, NY (Mudd, J. B. and Kozlowski T. T. (eds.)), pp. 57–87.
18. Davis, D. D. (1977). Responses of ponderosa pine primary needles to separate

and simultaneous ozone and PAN exposures. *Plant Disease Reptr.*, **61**: 640–44.

19. DAVIS, D. D. and R. G. WILHOUR. (1976). *Susceptibility of woody plants to sulfur dioxide and photochemical oxidants*, US Environmental Protection Agency, Pub. 600/3–76–102.
20. ERNST, W. (1976). Physiological and biochemical aspects of metal tolerance. In: *Effects of Air Pollutants on Plants* (Mansfield, T. A. (Ed.)), Cambridge Univ. Press, 115–33.
21. FEDER, W. A. (1978). Plants as bioassay systems for monitoring atmospheric pollutants. *Env. Health Pers.*, **27**: 139–47.
22. FERRY, B. W., M. S. BADDELEY, and D. L. HAWKSWORTH. (1973). *Air Pollution and Lichens*, Athlone Press, London.
23. HANAN, J. J. (1973). Ethylene dosages in Denver and marketability of cut-flower carnations. *J. Air Poll. Contr. Assoc.*, **23**: 522–4.
24. HAWKSWORTH, D. L. (1971). Lichens as litmus for air pollution: A historical review. *Int. J. of Environ. Studies*, **1**: 281–96.
25. HAWKSWORTH, D. L. (1974). Literature on air pollution and lichens. *Lichenologist*, **6**: 122–5.
26. HEATH, R. L. (1975). Ozone. In: *Responses of Plants to Air Pollution.* (J. B. Mudd and Kozlowski T. T. (eds.)), Academic Press, N. Y.
27. HECK, W. W. (1968). Factors influencing expression of oxidant damage to plants. *Ann. Rev. Phytopathology*, **6**: 165–88.
28. HECK, W. W., F. L. FOX, C. S. BRANDT, and J. A. DUNNING. (1969). *Tobacco, a sensitive monitor for photochemical air pollution*, US National Air Poll. Contr. Admin., Pub. No. AP-55.
29. HECK, W. W., R. H. DAINES, and I. J. HINDAWI. (1970). Other phytotoxic pollutants. In: *Recognition of Air Pollution Injury to Vegetation: A Pictorial Atlas.* (Jacobson, J. S. and Hill A. C. (eds.)), Air Poll. Control Assoc., Pittsburgh, Pa., pp. F1–F24.
30. HECK, W. W., A. S. HEAGLE, and E. B. COWLING. (1977). Air pollution: Impact on plants. In: *Proc. 32nd Meeting Soil Conservation Soc. of Am.*, 193–203.
31. HILL, A. C., H. E. HEGGESTAD, and S. N. LINZON. (1970). Ozone. In: *Recognition of Air Pollution Injury to Vegetation: A Pictorial Atlas.* (Jacobson, J. S. and Hill, A. C. (eds.)), Air pollution Control Association, Pittsburgh, Pa., pp. B1–B22.
32. JACOBSON, J. S. and L. J. COLAVITO. (1976). The combined effect of sulfur dioxide and ozone on bean and tobacco plants. *Environ. and Exptl. Botany*, **16**: 277–85.
33. JACOBSON, J. S. and W. A. FEDER. (1974). *A Regional Network for Environmental Monitoring: Atmospheric Oxidant Concentrations and Foliar Injury to Tobacco Indicator Plants in the Eastern United States.* Bulletin No. 604 of the Massachusetts Agricultural Experiment Station, University of Massachusetts, Amherst. 31 pp.
34. LEBLANC. F., and D. N. RAO. (1973). Effects of sulphur dioxide on lichen and moss transplants. *Ecology*, **54**: 612–17.
35. LACASSE, N. L., and M. TRESHOW (eds.) (1976). *Diagnosing Vegetation Injury Caused by Air Pollution*, US Environmental Protection Agency Handbook.

36. LERMAN, S. L., and E. F. DARLEY. (1975). Particulates. In: *Responses of Plants to Air Pollution*. (Mudd, J. B. and Kozlowski T. T (eds.)), Academic Press, NY.
37. LEWIS, E., and E. BRENNAN. (1978). Ozone and sulfur dioxide mixtures cause a PAN-type injury to petunia. *Phytopathology*, **68**: 1011–14.
38. LINZON, S. N. (1969). Symptomatology of sulfur dioxide injury on vegetation. In: *Handbook of Effects Assessment: Vegetation Damage*. (Lacasse, N. L. and W. J. Moroz (eds.)), Penn. State Univ., pp. VIII–1 to VIII–13.
39. LITTLE. P. (1973). A study of heavy metal contamination of leaf surfaces. *Environ. Pollut.*, **5**: 159–172.
40. LITTLE, P., and M. H. MARTIN. (1972). A survey of zinc, lead, and cadmium in soil and natural vegetation around a smelting complex. *Environ. Pollut.* **3**: 241–54.
41. MCCUNE, D. C., A. E. HITCHCOCK, J. S. JACOBSON, and L. H. WEINSTEIN. (1965). Fluoride accumulation and growth of plants exposed to particulate cryolite in the atmosphere. *Contrib. Boyce Thompson Inst.*, **23**: 1–12.
42. MALONE, C., D. E. KOEPPE, and R. J. MILLER. (1974). Localization of lead accumulated by corn plants. *Plant Physiology*, **53**: 388–394.
43. MANNING, W. J. (1971). Effects of limestone dust on leaf condition, foliar disease incidence, and leaf surface microflora of native plants. *Environ. Pollut.* **2**: 67–74.
44. MUDD, J. B. (1975). Sulfur dioxide. In: *Responses of Plants to Air Pollution*. (Mudd, J. B. and Kozlowski T. T. (eds.)), Academic Press, NY, 9–22.
45. MUDD, J. B. (1975). Peroxyacyl nitrates. In: *Responses of Plants to Air Pollution*. (Mudd, J. B. and Kozlowski T. T. (eds.)), Academic Press, NY, 97–116.
46. NIEBOER, E., D. H. S. RICHARDSON, K. J. PUCKETT, and F. D. TOMASSINI. (1976). The phytotoxicity of sulfur dioxide in relation to measurable responses in lichens. In: *Effects of Air Pollutants on Plants*. (Mansfield, T. A. (Ed.)), Cambridge Univ. Press, London, 61–85.
47. ORMROD, D. P., N. O. ADEDIPE, and D. J. BALLANTYNE. (1976). Air pollution injury to horticultural plants: A review. *Hort. Abstracts* **46**: 241–246.
48. ORMROD, D. P. (1978). *Pollution in Horticulture*. Elsevier Scient. Pub. Co., Amsterdam, 260 pp.
49. REINERT, R. A. (1975). Monitoring, detecting, and effects of air pollutants on horticultural crops: Sensitivity of genera and species. *Hort. Sci.*, **10**: 7–12.
50. REINERT, R. A., A. S. HEAGLE, and W. W. HECK. (1975). Plant responses to pollutant combinations. In: *Responses of Plants to Air Pollution*. (Mudd, J. B. and Kozlowski T. T. (eds.)), Academic Press, NY, 159–75.
51. RICHARDS, B. L., J. T. MIDDLETON, and W. B. HEWITT. (1958). Air pollution with relation to agronomic crops: V. Oxidant stipple of grapes. *Agron. Jour.*, **50**: 559–61.
52. RICKS, G. R., and R. J. H. WILLIAMS. (1974). Effects of atmospheric pollution on deciduous woodland. Part 2: Effects of particulate matter upon stomatal diffusion resistance in leaves of *Quercus petraea* (Mattuschka) Leibl. *Environ. Pollut.*, **6**: 87–109.
53. SPIERINGS, F. H. F. G. (1967). Chronic dislocation of leaf tips of gladiolus and its relation to the hydrogen fluoride content of the air and the fluorine content of the leaves. *Neth. J. Plant Pathol.*, **73**: 25–28.

54. Spierings, F. H. F. G. (1971). Influence of fumigations with NO_2 on growth and yield of tomato plants. *Neth. J. Plant Pathol.*, **77**: 194–200.
55. Taylor, O. C. (1969). Importance of peroxyacetyl nitrate (PAN) as a phytotoxic air pollutant. *J. Air Poll. Contr. Assoc.*, **19**: 347–51.
56. Taylor, O. C. (1970). Effects of photochemical oxidants on vegetation in urban and surrounding areas. In: *Impact of Air Pollution Conference.* (Linzon, S. N. (ed.)), Air Poll. Contr. Assoc., Pittsburgh, Pa.
57. Taylor, O. C. and D. C. MacLean. (1970). Nitrogen oxides and the peroxyacyl nitrates. In: *Recognition of Air Pollution Injury to Vegetation: A Pictorial Atlas.* (Jacobson, J. S. and Hill A. C. (eds.)), Air Poll. Contr. Assoc., Pittsburgh, Pa., E1–E5.
58. Temple, P. J., and S. N. Linzon. (1976). Boron as a phytotoxic air pollutant. *J. Air Poll. Contr. Assoc.*, **26**: 498–99.
59. Temple, P. J. and R. A. Richards. (1978). Effects of atmospheric deposition of sodium sulfate on bean and tomato plants. *Bull. Environ. Contam. and Toxicol.*, **19**: 257–63.
60. Thompson, C. R., G. Kats, and E. G. Hensel. (1971). Effects of ambient levels of NO_2 on navel oranges. *Environ. Sci. and Technol.*, **5**: 1017–19.
61. Tingey, D. T., R. A. Reinert, J. A. Dunning, and W. W. Heck. (1971). Vegetation injury from the interaction of nitrogen dioxide and sulfur dioxide. *Phytopathology*, **61**: 1506–11.
62. Tornebene, T. G., N. L. Gale, D. E. Koeppe, R. L. Zimdahl, and R. M. Forbes. (1977). Effects on microorganisms, plants, and animals. In: *Lead in the Environment*, (Boggess, W. R. and B. G. Wixson (eds.)), US National Science Foundation Report NSF/RA-770214, 93–98.
63. Treshow, M. (1971). Fluorides as air pollutants affecting plants. *Ann. Rev. Phytopathol.*, **9**: 22–43.
64. Treshow, M., and M. R. Pack. (1970). Fluoride. In: *Recognition of Air Pollution Injury to Vegetation: A Pictorial Atlas.* (Jacobson, J. S. and Hill A. C. (eds.)), Air Poll. Contr. Assoc., Pittsburgh, Pa., pp. D1–D17.
65. Troiano, J. J., and I. A. Leone. (1977). Changes in growth rate and nitrogen content of tomato plants after exposure to NO_2. *Phytopathology*, **67**: 1130–3.
66. Van Haut, H. and H. Stratmann. (1970). Farbafelatlas über Schwefeldioxid-Wirkungen an Pflanzen. Verlag W. Girardet, Essen, Germany (F.R.). 206 pp.
67. Weinstein, L. H., and D. C. McCune. (1970). Field surveys, vegetation sampling, and air and vegetation monitoring. In: *Recognition of Air Pollution Injury to Vegetation: A Pictorial Atlas.* (Jacobson, J. S. and Hill A. C. (eds.)), Air Pollution Control Association, Pittsburgh, Pa., G1–G4.
68. Weinstein, L. H., and D. C. McCune. (1970). Effects of fluorides on vegetation. In: *Impact of Air Pollution Conference.* (Linzon, S. M. (ed.)), Air Poll. Contr. Assoc., Pittsburgh, Pa., 81–101.
69. Wu, L., and A. D. Bradshaw. (1972). Aerial pollution and the rapid evolution of copper tolerance. *Nature, London*, **238**: 167–169.
70. Zeevaart, A. J. (1976). Some effects of fumigating plants for short periods with NO_2. *Environ. Pollut.*, **11**: 97–108.
71. Zimdahl, R. L. (1976). Entry and movement in vegetation of lead derived from air and soil sources. *J. Air Poll. Control Assoc.*, **26**: 655–660.

CHAPTER 3

Plants as Indicators and Biomonitors

I. INTRODUCTION

An indicator plant is one which exhibits injury symptoms when exposed to phytotoxic concentrations of pollutant or pollutant mixture. A plant indicator is a chemical sensor which can detect the presence of a pollutant in the air. Monitoring connotes something more quantitative than indicating. A plant monitor must be an indicator, but it must also help answer the question: How much?

Pollution injury to vegetation has been thoroughly documented[1,5,6,62–64] and is well illustrated in several excellent atlases.[31,35,65] Plants can also act as indicators by accumulating the pollutant or some metabolic product of the plant/pollutant interaction in their tissues.[33,34] Trace metals, such as lead or cadmium, and gaseous pollutants like hydrogen fluoride (HF) or sulfate (SO_4), from SO_2 exposure, are examples of this phenomenon.[18–21] Plants may also exhibit altered growth rates, changes in rates of maturation, reduction of flower, fruit and seed formation, alterations in the reproductive process and, ultimately, depression of productivity and yield.[2,9,23,24,44,49,50] Any or all of these parameters could be used to indicate the presence of pollutants in ambient air, provided that testing under controlled conditions has been carried out so as to clearly relate the symptoms or change in plant performance to a particular pollutant or pollutant mixture. It is clear, for example, after much research, that Bel-W 3 tobacco is very sensitive to ozone and yields a definitive succession of symptoms when exposed to known doses of ozone.[26,27,42] It has also been demonstrated that fruit set and yield of tomato cv. 'Tiny Tim' are significantly reduced when this plant is chronically exposed to low ozone concentrations.[9,44] Similarly, soybean cultivars exposed to known doses of SO_2 exhibit recognizable injury symptoms and alterations in growth patterns and yield.[24] Many other species and cultivars of plants have been shown to act as indicators of air pollutants.

Other plant species, especially lichens and mosses, have been shown to act as collectors of air pollutants, especially trace metals, which these plants can accumulate to levels well above ambient concentrations.[1,4,8,12–16,22,30,38,45,51,56–58,60,61]

The appearance of 'typical' injury symptoms, or the presence of measureable amounts of metabolites or trace metals, *indicates* that a pollutant, or mixture of pollutants, is present in the ambient air surrounding that particular plant. A fundamental question remains—how to find a way to quantify these observations so that the *indicator* plant becomes a *monitor*. It is our purpose here to show how this may be done and to illustrate this with some examples.

Let us examine the problem more closely. If an air pollutant can be precisely measured by instrumentation why try to do the same thing with plants, which are variable, living creatures? One answer is cost. An accurate ozone metering instrument, for example, may cost upwards of $5000. It needs power to run it and constant attention to its calibration and proper functioning. Sulfur dioxide monitors cost at least as much and a Panalyser may cost even more. High volume air samplers, used in collecting airborne particulates, are also expensive and require maintenance. All of these are highly sensitive, delicate instruments, not well suited for functioning in the weather encountered at remote sites. These instruments are essential research tools, but the cost of a well-designed monitoring network using such instruments usually far exceeds the resources of most agencies charged with pollutant monitoring or abatement.

Another factor which militates against the use of these expensive, sensitive instruments at remote sites is that the instrument response is considerably affected by environmental factors over which there is no control. If control is achieved, it is at great expense.

Finally, as the state of the art improves, data gathered using a particular instrument may no longer be considered valid by the major regulatory agencies responsible for establishing criteria and standards.

What does the green plant offer? Per unit, it is inexpensive, easily reproduced, readily multiplied and capable of several different response modes from which the observer can choose one or more best suited to a particular study. There is a choice between short-lived (herbaceous) plants which must be replaced every season, or several times during one growing season, and woody plants (trees and shrubs) which can be set out in the desired locations and will continue to grow and *indicate* for many seasons with relatively little attention.

But we want our indicator to monitor—to quantify—the quality of the air. This means that we must discover and utilize some relationship between plant response and ambient pollutant concentration.

There are three general ways in which this can be done. The first is to integrate the degree of pollutant-incited injury with known ambient pollutant concentrations. The second is to use the plant as a living collector and the third is to measure the amount of pollutant or pollutant-related metabolite which appears in the plant tissues after exposure to the pollutant and to relate that quantity to the amount of pollutant in the ambient air.[10,11]

Plants growing naturally or planted in the field are exposed to the entire range of pollutants present in ambient air. As the plants grow, they integrate all those stresses and the result is a normal plant or one which is shorter, has fewer flowers, fails to set seeds, has spotted leaves, or discolored needles, or does not grow at all.[3,5-7,9-11,23,24,27,28,44,48,50,54,62,64]

Inherent genetic variability allows a range of responses of species and cultivars to pollutant stresses. One plant species may react to one pollutant in a characteristic way. Some plant species react to two or more pollutants, some plants do not respond at all, or at best minimally, to the same pollutant stress. Cultivars within species show great variability in their response to a pollutant or pollutant mixtures.[3,10,28,29,44,54] It is therefore possible to select the particular species or cultivar which will yield the desired response to a given pollutant. This variable response to pollutant stresses is present in mosses, ferns, gymnosperms and angiosperms, so that selections amongst plant species and/or cultivars in any of these groups can be made with the intent to use the plants as bioindicators and/or biomonitors.

Natural populations, plus a breeding and selection effort targeted to produce pollutant-tolerant and susceptible lines of plants, should yield the plant types useful for monitoring air pollutants.

Diligence and time will produce the necessary plant materials but more work is needed to establish a monitoring network based on the responses of living plants. Plant injury symptoms, altered growth and reproductive patterns, changes in yield and/or productivity and changes in species distribution can be used singly or in combination as monitoring devices. However, all these responses are, in turn, greatly influenced by plant age, environmental parameters and cultural conditions.[25] Soil type, soil moisture, nutritional status, relative humidity, topography and meteorological conditions all affect the manner in which a particular plant responds to a particular concentration or dose of any pollutant or mixture of pollutants. Variability within selected plant populations dictates that sizeable numbers of plants be used at a selected monitoring site and that the plants be arrayed spatially to get maximum exposure of the plants to pollutant-bearing winds.

Tolerant and resistant cultivars of the same plant species can be set out at a given site to respond to a single pollutant or a whole plant garden can be utilized by setting out several different plant species each susceptible to a different pollutant or pollutant mixture.[3,10,11,54]

Plants can be set out in pots, baskets or buckets or planted directly into the soil at a prepared site. Oshima[48] included a passive automatic watering device for his 'air monitoring biological indicator system' so that the plants could remain unattended for a week in the field under dry California conditions. To minimize variation, it is important to use the same soil mix and/or soil type for all plants in the network system. Plants set directly in the ground should also have the hole filled with a known soil mix.

II. RELATING PLANT RESPONSES TO POLLUTANT CONCENTRATIONS

It is absolutely essential that plant responses to a particular pollutant or pollutant mixture be clearly defined and understood before setting out a field experiment. This entails considerable controlled plant screening to establish dose–response relationships. Whether the response is leaf injury, growth, yield or bioaccumulation, sufficient studies should have been performed to understand how the plant responds to doses of a single pollutant or pollutant mixtures. It has taken many years to establish the dosage–response curve for Bel-W 3 tobacco and ozone,[26–28,42] for chlorotic dwarf of eastern white pine[3,32,54] or soybean and SO_2 or O_3[24] and *Hypnum* spp. and several heavy metals, to mention just a few examples.[8,15,17,51,53,56,57]

Leaf injury can be measured by a photo reference system utilizing direct comparisons of injured leaves with reference photographs taken of leaves of plants exposed under controlled conditions to known pollutant concentrations.[48] If statistically significant numbers of plants are arranged in some kind of grid covering the test site, then leaf injury data can be displayed as number of leaves affected, severity of leaf injury, number of lesions per unit area, etc. and curves drawn relating leaf injury to time and pollutant dose. The slopes of these curves can be compared with slopes of dosage/response curves developed earlier under controlled conditions. The air quality can then be quantified with regard to a particular pollutant mixture with which the controlled curves were generated.[10,11] The plants will tell the investigator that the quality of the ambient air in that particular environment during that specific time period has been good, fair or poor, and should also identify the pollutant or pollutant mixture. The terms 'good', 'fair' and 'poor' can be related to the dosage/response curves generated earlier under controlled conditions for that plant/pollutant interaction, thus monitoring the range of concentrations of the pollutants during the episode described or measured.

III. SELECTING PLANT MATERIAL

General criteria used in selecting plant material for biomonitoring of air pollutants can now be discussed. Certain characteristics are essential in selecting which plants to use as biomonitors. The plant must first exhibit a clear cut pollutant response. This could include visible injury symptoms, growth or morphological changes, changes in flower, fruit or seed production or changes in productivity or yield. The ability to reproduce this response must be assured by the use of a large, uniform, readily available supply of seed or other plant propagules. The plants should be easy to grow and manage using normal cultural practices. Plants that present pest and disease control problems should not be considered as suitable candidates for biomonitors. A single source of seeds or propagules should be used to ensure uniformity of results in studies carried out by different investigators in different geographical locations.

An understanding of the genetic make-up of the plant material is also most useful. Genetic markers should be noted to aid the worker in identifying his plant material and assuring that the response is being developed by the material that has been designated. For annual plants, several seasons' plantings should be made to observe if the plant material selected behaves locally as described by persons working in other regions. A uniform soil mix is also necessary.

IV. EVALUATING PLANT RESPONSES

Rating systems for quantifying data vary, depending upon the plant material, pollutant and the parameters to be measured. Leaf injury of herbaceous plants such as beans or tobacco is usually measured by a visual estimation of the per cent leaf area which is bronzed, stippled, flecked or otherwise injured. Bean plants are only useful until the third set of trifoliate leaves appears.[11] Tobacco plants can be used until they flower and even beyond the flowering stage if the flowers are removed as they form. Gladiolus plants can be used for only one season, while white pine and grape can be used for as long as they live.[6,10,11,25,27,28,29,33,42,47] A typical rating scale for bean plants is shown in Table 7.

Data can be assembled for: (1) per cent of leaf area injured, (2) new injury on each plant for any given time period or (3) leaf area. In the case of needle injury data should include: (1) needle length, (2) needle color,

TABLE 7
A RATING SYSTEM FOR EVALUATING RESPONSE OF GARDEN BEAN (*Phaseolus vulgaris* L.) TO AMBIENT OZONE

	Rating systems	
Rating of injury	*Injury severity index*	*Per cent leaf injury*
None	0	0
Slight	1	1–25
Moderate	2	26–50
Moderate–severe	3	51–75
Severe	4	76–99
Complete	5	100

Example—Primary leaves

Date (days from seed)	*Injury severity index*	*Per cent leaf injury*
7	0	0
11	1	15
15	2	40 (+25)
20	4	80 (+40)

Injury is rated by visual estimation of per cent leaf area bronzed or stippled. Cumulative readings can be made at 3–5 day intervals until the leaf falls. (See examples above). Plants should no longer be used after the third set of trifoliate leaves has expanded as it becomes too difficult and time consuming to keep track of the leaves on a cumulative basis.

Developed by W. J. Manning, and adapted from: Feder and Manning.[11]

(3) needle shape, (4) needle age, and (5) per cent needles injured per twig.

When growth and productivity data are used to measure response, the information to be collected should include: (1) rate of growth, (2) number of leaves and/or leaf area, (3) time to bud formation, (4) time to flowering, (5) ratio of number of buds to number of flowers, (6) ratio of flowers/fruits, (7) number of seeds per fruit, (8) ratio of shoots/roots and (9) total yield (or biomass). For trees it is important to know: (1) twig number, (2) twig length, (3) twig diameter, (4) trunk diameter at a given point above the ground level, (5) growth rate of the trunk, (6) leaf or needle size and/or area, (7) fruit or cone set and (8) seed set.

Where pollutant uptake is to be used to measure pollution levels either the pollutant itself or a metabolite of the pollutant is measured. Tissue sulfate can be related to SO_2, tissue fluoride to ambient HF.[10,14,18–21,30,34,37,43] Equations can be derived to relate tissue levels to ambient levels of the pollutant. Careful standardization of tissue collection and processing and analytical techniques is essential to ensure that results obtained by different investigators are comparable.

Plants serving as living collectors are particularly useful in monitoring heavy metals. *Hypnum cupressiforme*, a moss, is capable of taking up heavy metals such as zinc, lead, cadmium, nickel, copper and magnesium. The metals are not only collected on the moss leaves, but are also absorbed by the plant body and accumulated in the tissues. By collecting the plants, drying and weighing them, and then subjecting the dried tissues to chemical analysis, the amount of metal uptake can be calculated.[17] By altering the time intervals between collections, the tissue metal content can be related to the concentration of metal, or metals, in the ambient air, Mosses can be taken from natural habitats, or they can be reared in a 'clean' environment and then hung at selected test sites to be collected and analysed at a later date.[8,15,51,53,56,57]

Lichen species can be used to monitor SO_2.[12,61] SO_2 accumulation by lichens is species dependent. By a combination of instrument monitoring and observation of lichen survival and speciation it is possible to correlate lichen growth and/or survival with ambient concentrations of SO_2.[14,16,38,58] Lichens can be collected from uncontaminated sources, grown under controlled conditions, and pieces of known dimensions can be displayed at test sites where growth can be carefully observed. This allows for the establishment of lichen networks covering large areas surrounding an active SO_2 source. The growth rate and the color of the lichen indicate the presence or absence of SO_2 and its approximate concentration in air masses passing over the test sites.[13,37,45,60] Once a clean, unexposed lichen supply is developed the system is simple and reliable and inexpensive to set out and manage. Such systems have been used to monitor SO_2 in England,[14,22,52] Ireland,[12,13] France,[45] Canada,[36-38] Sweden[60] and Long Island, USA.[4]

The material presented here comprises our rationale for suggesting that plants, including mosses, lichens, gymnosperms and angiosperms, can be useful as biomonitors for studying and quantifying air pollutants present in ambient air. An attempt is made to distinguish between bioindicators and biomonitors and several examples are given illustrating how one might use plants to monitor both gaseous and particulate pollutants. Suggestions as to the types of data that might be collected and how such data could be handled are also mentioned. Detailed methodology for dealing with specific pollutants will be dealt with in the chapters that follow.

REFERENCES

1. BARKMAN, J. J. (1968). The influence of air pollution on bryophytes and lichens. *First European Congr. Influence Air Pollution on Plants and Animals., Wageningen, Netherlands*, pp. 197–209.
2. BENNET, J. P. and R. J. OSHIMA. (1976). Carrot injury and yield response to ozone. *J. Amer. Soc. Hort. Sci.* **101**: 638–9.
3. BERRY, C. R. (1964). Eastern white pine: A tool to detect air pollution. *Southern Lumberman.*
4. BRODO, I. M. (1966). Lichen growth and cities: A study on Long Island, New York. *Bryologist*, **69**: 427–49.
5. DAINES, R. H. (1968). Sulfur dioxide and plant response. *J. Occup. Med.* **19**: 516–24.
6. DARLEY, E. F. (1960). Use of plants for air pollution monitoring. *J. Air Poll. Control Assoc.* **10**: 198–9.
7. DARLEY, E. F. (1966). Studies on the effect of cement-kiln dust on vegetation. *J. Air Poll. Control Assoc.*, **16**: 145–150.
8. ELLISON, G., J. NEWHAM, M. J. PINCHIN, and I. THOMPSON. (1976). Heavy metal content of moss in the region of Consett (Northeast England). *Environ. Pollution*, **11**: 167–74.
9. FEDER, W. A. (1977). Adverse effects of chronic low level ozone exposure to tomato fruit set and yield. In: *Proceedings of the 100th Anniv. Cottrell Symposium, Calif. State Univ., Stanislaus.* (Perona, M. (ed.)).
10. FEDER, W. A. (1978). Plants as bioassay systems for monitoring atmospheric pollutants. *Environmental Health Perspectives*, **27**: 139–47.
11. FEDER, W. A. and W. J. MANNING. (1979). Living plants as indicators and monitors. In: *Handbook of Methodology for the Assessment of Air Pollution Effects on Vegetation.* (Heck W. W., Krupa, S. V. and Linzon, S. N. (eds.)), Air Pollution Control Association, Pittsburgh, PA. pp. 9–1 to 9–14.
12. FENTON, A. F. (1950). Lichens as indicators of atmospheric pollution. *Irish Nat. J.* **13**: 153–9.

13. Fenton, A. F. (1964). Atmospheric pollution of Belfast and its relationship to the lichen flora. *Irish Nat. J.*, **14**: 237.
14. Gilbert, O. L. (1968). The effect of SO_2 on lichens and bryophytes around Newcastle-on-Tyne. *Proc. First European Congr. Influence Air Pollution Plants and Animals, Wageningen, Netherlands*, pp. 223–35.
15. Gilbert, O. L. (1968). Bryophytes as indicators of air pollution in the Tyne Valley. *New Phytol.* **67**: 15–30.
16. Gilbert, O. L. (1970). Further studies of the effect of sulphur dioxide on lichens and bryophytes. *New Phytol.*, **69**(3): 605–27.
17. Goodman, G. T. and T. M. Roberts. (1971). Plants and soils as indicators of metals in the air. *Nature (London)* **231**: 287–92.
18. Guderian, R. (1970). Untersuchungen ueber quantitative Beziehungen zwischen dem Schwefelgehalt von Pflanzen und dem Schwefeldioxidgehalt der Luft. *Z. Pflanzenkrankh. Pflanzenschutz*, **77**(4/5): 200–20.
19. Guderian, R. (1970). Untersuchungen ueber quantitative Beziehungen zwischen dem Schwefelgehalt von Pflanzen und dem Schwefeldioxidgehalt der Luft. Teil 2. Tagesgang im Schwefelgehalt bei unbeeinflussten und begasten Pflanzen. *Z. Pflanzenkrankh. Pflanzenschutz*, **77**: 289–308.
20. Guderian. R. (1970). Untersuchungen ueber quantitative Beziehungen zwischen dem Schwefelgehalt von Pflanzen und dem Schwefeldioxidgehalt der Luft. Teil 3. Schwefelgehalt geschaedigter and ungeschaedigter Blatteile. *Z. Pflanzenkrankh. Pflanzenschutz*, **77**: 387.
21. Guderian, R., H. Van Haut and H. Stratmann, (1971). Pflanzenschaedigende Fluorwasserstoff-Konzentrationen. *Umschau (Frankfort)*, **71**(21): 777.
22. Hawksworth, D. L., and F. Rose. (1970). Qualitative scale for estimating sulfur dioxide air pollution in England and Wales using epiphytic lichens. *Nature (London)*, **227**: 145–148.
23. Heagle, A. S., D. E. Body and E. K. Pounds. (1972). Effect of ozone on yield of sweet corn. *Phytopathology*, **62**: 683–7.
24. Heagle, A. S., D. E. Body, and G. E. Neely. (1974). Injury and yield responses of soybean to chronic doses of ozone and sulfur dioxide in the field. *Phytopathology*, **64**: 132–6.
25. Heck, W. W. (1968). Factors influencing expression of oxidant damage to plants. *Ann. Rev. Phytopathol.*, **6**: 165–88.
26. Heck, W. W., J. A. Dunning, and I. J. Hindawi. (1966). Ozone: Nonlinear relation of dose and injury in plants. *Science* **151**: 577–8.
27. Heck, W. W. and A. S. Heagle. (1970). Measurement of photochemical air pollution with a sensitive monitoring plant. *J. Air Poll. Control Assoc.*, **20**: 97–9.
28. Heggestad, H. E. and H. A. Menser. (1962). Leaf spot-sensitive tobacco strain Bel-W 3, a biological indicator of the air pollutant ozone. *Phytopathology*, **52**: 735.
29. Heggestad, H. E. and E. F. Darley. (1968). Plants as indicators of the air pollutants ozone and PAN. *Proc. First European Congr., Influence Air Pollution, Plants and Animals, Wageningen, Netherlands*. p. 329.
30. Hill, D. J. (1971). Experimental study of the effect of sulphite on lichens with reference to atmospheric pollution. *New Phytol.* **70**: 831–6.

31. HINDAWI, I. J. (1970). *Air pollution injury to vegetation. Natl. Air Pollution Control Assoc. Publ. No. AP-71, Raleigh*, NC, 44pp.
32. HOUSTON, D. B. (1974). Response of selected *Pinus Strobus* L. clones to fumigations with sulfur dioxide and ozone. *Can. J. Forest Res.*, **4**: 65–68.
33. JACOBSON, J. S., L. H. WEINSTEIN, D. C. MCCUNE and A. E. HITCHCOCK. (1966). The accumulation to fluorine by plants. *J. Air Pollution Control Assoc.* **16**: 412–17.
34. JACOBSON, J. S., D. C. MCCUNE, L. H. WEINSTEIN, R. H. MANDL and A. E. HITCHCOCK. (1966). Studies on the measurement of fluoride in air and plant tissues by the Willard-Winter and semiautomated methods. *J. Air Poll. Control Assoc.*, **16**: 367–71.
35. JACOBSON, J. S. and A. C. HILL. (1970). *Recognition of Air Pollution Injury to Vegetation: A Pictorial Atlas.* Air Pollution Control Assoc., Pittsburgh, Pa.
36. LEBLANC, F. (1961). Influence de l'atmosphere polluee des grandes agglomerations urbaines sur les epiphytes corticoles. *Rev. Can Biol.*, **20**: 823–7.
37. LEBLANC, F. and D. N RAO. (1966). Reaction of several lichens and epiphytic mosses to sulfur dioxide in Sudbury, Ontario. *Bryologist*, **69**: 338–46.
38. LEBLANC, F. and J. DE SLOOVER. (1970). Relation between industrialization and the distribution and growth of epiphytic lichens and mosses in Montreal. *Can. J. Bot.*, **48**: 1485–96.
39. LITTLE, P. (1973). A study of heavy metal contamination of leaf surfaces. *Environ. Pollution*, **5**: 159–72.
40. LITTLE, P., and M. H. MARTIN. (1972). A survey of Zn, Pb and Cd in soil and natural vegetation around a smelting complex. *Environ. Pollution* **3**: 241–54.
41. LITTLE, P. and M. H. MARTIN. (1974). Biological monitoring of heavy metal pollution. *Environ. Pollution*, **6**: 1–19.
42. MACDOWALL, F. D. H., E. I. MUKAMMAI and A. F. W. COLE. (1964). Direct correlation of air-polluting ozone and tobacco weather fleck. *Can. J. Plant Sci.*, **44**: 410–17.
43. MCCUNE, D. C. and A. E. HITCHCOCK. (1970). *Fluoride in forage: Factors determining its accumulation from the atmosphere and concentration in the plant.* Preprint, Inter. Union of Air Pollut. Prevention Assns., 17 pp.
44. MANNING, W. J. and W. A. FEDER. (1976). Effects of ozone on economic plants. In: *Effects of Air Pollutants on Plants.* (Mansfield, T. A. (ed.)) Soc. for Experimental Biology Seminar Series, Vol. 1, Cambridge Univ. Press.
45. MARTIN, J. F. and F. JACQUARD. (1968). Influence of factory smoke on lichen distribution in the Romanche Valley (Isere). *Pollut. Atmos. (Paris)* **10**: 95–99.
46. MATUOKA, Y. (1971). Nosakubutsutai seibun kara mita taiki osen shitsu kakusan ni tsuite. *Kogai to Taisaku*, **7**: 21–35.
47. MENSER, H. A., H. E. HEGGESTAD, and O. E. STREET. (1963). Response of plants to air pollutants. II. Effects of ozone concentration and leaf maturity on injury to *Nicotiana tabacum. Phytopathology*, **53**: 1304–8.
48. OSHIMA, R. J. (1974). A viable system of biological indicators for monitoring air pollutants. *Air Poll. Control Assoc.*, **24**: 576–8.
49. OSHIMA, R. J., *et al.* (1975). Effect of ozone on the yield and plant biomass of a commercial variety of tomato. *J. Environ. Qual.*, **4**: 463–4.
50. OSHIMA, R. J., *et al.* (1976). Ozone dosage-crop loss function for alfalfa: A

standardized method for assessing crop losses from air pollutants. *J. Air Poll. Control Assoc.*, **26**: 861–5.

51. PILEGAARD, K. (1976). Tungmetaller i regnvanc, jord og planter omkring et kraftvaerk. *Danske Elvaekers Foreing. Driftstekniske Publikationer* **18**: 1–52.
52. PYATT, F. B. (1970). Lichens as indicators of air pollution in a steel production town in South Wales. *Environ. Pollut.* **1**: 45–56.
53. RASMUSSEN. L. (1977). Epiphytic bryophytes as indicators of the changes in background levels of airborne metals from 1951-1975. *Environ. Pollution* **14**: 37–45.
54. ROBERTS, B. R. (1976). The response of field-grown white pine seedlings to different sulphur dioxide environments. *Environ. Pollut.*, **11**: 175–80.
55. ROBERTS, T. M. (1972). Plants as monitors of airborne metal pollution. *J. Environ. Plant Pollut. Control* **1**: 43–54.
56. RUHLING, A. and G. TYLER. (1969). Ecology of heavy metals—A regional and historical study. *Bot. Notiser*, **122**: 248–259.
57. RUHLING, A., and G. TYLER. (1970). Sorption and retention of heavy metals in woodland moss *Hypnum splendens* (Hedw) Br. et Sch. *Oikos*, **21**: 92–97.
58. SCHOENBECK, H. (1969). A method for determining the biological effects of air pollution by transplanted lichens. *Staub* (Engl. transl. of *Staub, Reinhaltung Luft*) **29**: 17–21.
59. SKAWINA, T. *et al.* (1964). Zastosowanie testu roslinnego do oceny stopnia zanieczyscenia atmosferycznego dwutlenkien siarki. *Postepy Nauk Roin (Warsaw)*, **3**: 17–31.
60. SKYE, E. (1968). Lichens and air pollution. A study of cryptogamic epiphytes and environment in the Stockholm region. *Suecica*, **52**: 138.
61. SKYE E. (1969). The use of lichens as indicator and test organism for atmospheric pollution. *Nort. Hyg. Tidskr (Stockholm)*. No. 3: 115–34.
62. THOMAS, M. D. and R. H. HENDRICKS. (1956). Effects of air pollution on plants. In: *Air Pollution Handbook*, (Magill, P. L. *et al.* (eds.)), McGraw-Hill, New York, pp. 9–1 to 9–44.
63. THOMAS, M. D. (1958). Air pollution with relation to agronomic crops. I. General status of research on the effects of air pollution on plants. *Agron. J.* **50**: 545–550.
64. THOMAS, M. D. (1961). Effects of air pollution on plants. In: *Air Pollution*, World Health Organization, Geneva, pp. 233–78.
65. VAN HAUT, H. and H. STRATMANN. (1970). *Farbtafelatlas uber Schwefeldioxid-Wirkung an Pflanzen.* W. Girardet, Essen, West Germany., 206 pp.

CHAPTER 4

Biomonitoring Photochemical Oxidants

I. INTRODUCTION

The three major constituents of 'photochemical smog' are ozone (O_3), the peroxyacyl nitrates (PANs) and the oxides of nitrogen (NO_x). Four peroxyacyl nitrates have been identified: peroxyacetyl nitrate (PAN), peroxypropionyl nitrate (PPN), peroxybutyryl nitrate (PBN) and peroxyisobutyryl nitrate ($P_{iso}BN$). The only one in the series which has been extensively studied is PAN which is also the only constituent to occur in photochemical smog at phytotoxic concentrations.[70,98] For these reasons, it is the only one of the peroxy nitrate series which will be treated here. The oxides of nitrogen (NO_x) include nitric oxide (NO), nitrogen dioxide (NO_2) and nitrogen pentoxide (N_2O_5). Only NO and NO_2 are thought to be phytotoxic. They are relatively non-phytotoxic when compared with O_3 and PAN.[98]

Background levels are about 0·02–0·04 ppm for O_3, and less than 0·01 ppm for PAN. NO_x levels are highly variable, quite dependent on traffic level variations and may range from less than 0·10 to 0·30 ppm in urban areas.[81, 82] Ozone concentrations of 0·50 to 0·70 ppm have been measured for 1 h in California. Concentrations of PAN have reached 0·06 ppm for 2 h and NO_x levels may exceed 0·50 ppm for 1 h.[70,81,82]

Oxidants are all gaseous pollutants and their production, concentration and dispersion are weather-related.[31,44,46,50] Air stagnation allows the build up of oxidants.[70] Polluted air masses may move and some of the oxidants and/or their precursors may move great distances over both land and water.[20,21,50,76] Evidence is accumulating that oxidant concentrations may peak later in the day and fall much more slowly during the dark hours than was previously believed.[50,76] Concentrations exceeding 0·15 ppm O_3 have been recorded after midnight in Waltham,

Massachusetts, USA, where ozone is monitored continuously for the entire year. Ozone injury on vegetation was discovered on Nantucket Island, which is off the southern coast of Massachusetts, and 300–500 km northeast of the nearest land source of oxidants. Recording of an O_3 concentration of 0·22 ppm for over 3 h on that island seems clear evidence that oxidants are capable of being carried long distances, even over water, without diminishing their phytotoxic concentrations.[50,76]

II. BIOMONITORING SYSTEMS FOR OXIDANTS

In order to devise a monitoring network for O_3 and other oxidants we must consider a complex series of interrelated responses between the plant receptor, the pollutant and the plant environment.[56,57,116] Factors affecting plant response to oxidant pollutants include: (1) the genetically determined thresholds for injury and/or physiological responses (growth rate, flowering, fruit and seed set, yield etc.), (2) plant age or state of maturity of the plant or plant parts, (3) cultural conditions, including nutrient level, soil moisture, light, temperature, relative humidity and soil type, etc., (4) concentration of the pollutant and duration of the exposure period (dosage/response), high pollutant concentrations for short exposure periods (acute exposure), low pollutant concentrations for long or repeated short exposure periods (chronic exposure) and (5) other meteorological factors such as wind speed, wind direction, persistence of wind patterns, rainfall and cloud cover.[2,42–45,48,53–55,57,74,75,84,87,99,100–105,107,108,116] All of these factors are integrated by the plant which then responds in characteristic manner to a particular pollutant.[56] However, since the plant can act as an integrator of all these factors, a change in even a single factor may alter the plant's response to the pollutant.[40,57] Designing a biomonitoring system for measuring any of the oxidants demands consideration of—or at least an awareness of—all of the above-listed factors. Some of the factors are known, some can be measured and monitored and some must have been worked out beforehand so that the investigator has an understanding of how the plant material will behave under certain known conditions in order to examine and quantify the unknowns encountered in the field. In addition, the investigator must know how the plant responds to diseases and pests to which it is susceptible, so that biogenic symptomotology can be distinguished from injury or other effects due to exposure to an air pollutant.[44,46] This is not always possible, so great care must be taken to use plant material which does not suffer from what are called 'mimicking symptoms'.[70,81,82]

III. SELECTION OF PLANT MATERIAL

A plant biomonitor must be part of a genetically uniform population whose members always react in a characteristic manner to a pollutant stress. The plant/pollutant reaction must be clear-cut.[1,3,11,59,60,63–66,70] The ability to reproduce this response must be assured by the availability of a large, uniform supply of like propagules (seeds or cuttings). Plants should be easy to grow and manage using normal cultural practices and should present minimal problems of disease and pest control. A single source of seeds and stock plants should be used to further ensure uniformity of plant and plant response. The genetic constitution of the material should be well studied. Genetic markers should be noted and used where possible to assure the worker of the identity of the material. There is a wide diversity of response within plant species to a particular pollutant stress. This variability in response is in part genetically controlled.[3,5,7,9,10,18,19,30,31,42,46,53,63,68,72,88] While this poses certain difficulties in obtaining uniform, reproducible plant/pollutant interactions, it also presents the worker with a unique opportunity to tailor plant material to give quite specific responses to particular pollutant stresses.

IV. BIOMONITORING OZONE

A large number of native and cultivated perennial and annual plants have one or more species that are known to be sensitive to O_3.[100–102, 112–114, 126–128, 133, 136, 137, 140–144] A few of these have been used to develop biomonitoring schemes for ambient O_3.

A. Tobacco

'Weather fleck' disease of tobacco (*Nicotiana tabacum* L.) is caused by ambient O_3[65,66,70,86] and is a specific response to O_3[91–96] (Fig. 19). The cultivar Bel-W3 is extremely sensitive to low concentrations of O_3.[64] The degree of leaf injury can be correlated with ambient O_3 concentrations.[85,86,94] Cultivar Bel-B is quite tolerant of O_3 and can be used for comparative studies with Bel-W3.[95]

FIG. 19. Extensive ozone-induced weather-fleck on Bel-W3 tobacco leaf (left) compared with a non-injured leaf (right) from a plant grown in a charcoal-filtered air chamber. (Courtesy: W. A. Feder.)

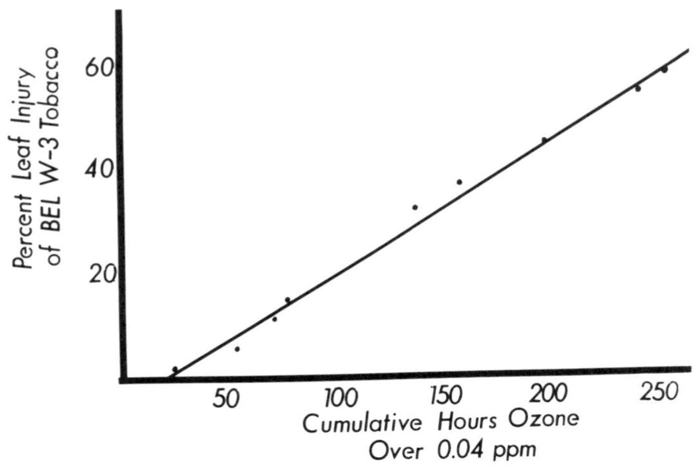

FIG. 20. Example of a linear dose/response curve for O_3 and Bel-W3 tobacco. (Courtesy: W. A. Feder.)

Heck *et al.*[58] determined that, although there was a positive correlation between degree of leaf injury and ambient O_3 concentrations, the relationship appeared to be non-linear.[58] Feder,[46] Feder and Manning,[47] and Feder *et al.*[50] however, demonstrated that a linear relationship could be developed if the number of plants per site were increased, the test site area were increased and a grid pattern for plant test sites were devised. An example of a linear dose/response curve from a field study with expanded parameters is shown in Fig. 20.

Plants of both cultivars can be set out, observed and the degree of injury sustained by plants of each cultivar can be recorded. Injury on Bel-W3 can be related to instrument-measured concentration of O_3 and a mathematical relationship between plant injury and ambient O_3 concentrations can be derived. The difference in the degree of injury sustained by the O_3-sensitive cultivar Bel-W3 and the O_3-tolerant cultivar Bel-B is a measure of air quality, as it relates to O_3.[47]

1. Growing plants

Seeds from a single source of both Bel-W3 and Bel-B are sown in vermiculite and watered with half-strength Hoagland's complete nutrient solution. Germination and growth take place in constant light at about 24°C in a charcoal-filtered chamber (to remove oxidants and SO_2). Seedlings are transplanted to small 'Jiffy' paper peat pots filled with a soil mixture consisting of peat/perlite (1 : 1) plus lime and gypsum (see Table 8 for details). Plants are watered with half-strength Hoagland's solution until they reach the four-leaf stage. They are then transplanted into 10-in pots filled with the same planting medium without removing the transplants from the paper pots. Watering is then done using tap water and the plants are fed with 15–15–15 liquid fertilizer once a week. Great care must be exercised to keep plants free of tobacco mosaic virus and insect pests, like white fly. It is imperative that the plants are reared in charcoal-filtered, O_3-free air until they are ready for setting out in the field. The smallest seedlings will show injury if they are exposed to extremely low O_3 levels. Plants should be set outdoors about 5 weeks after being transplanted into the Jiffy pots. Timing depends upon local conditions. Plants should have at least four true leaves before they are set out as monitors. They should be uniform in color and size and free of disease or pests.

Plants are set out in 2-gall plastic pails equipped with drainage holes. Pails are filled with a soil/peat/perlite mixture (Table 8). Enough of the soil mixture should be prepared to handle two plantings of thirty-six plants each or a total of seventy-two plants. This results in a total of 144 plants,

TABLE 8

PROCEDURES FOR MAKING SOIL MIXTURES USED IN GROWING TOBACCO PLANTS USED IN O_3 BIOMONITORING SCHEMES

A. *Procedure for making peat–perlite mixture for growing plants*
   (1) Mix in cement mixer.
   (2) Two pails (10 qts each) of peat moss (level full).
   (3) Two pails (10 qts each) of perlite (level full). To each pail of perlite add 2 liters of distilled or deionized water until all perlite is wet. This equals 4 liters per mix.
   (4) When peat moss and perlite are well mixed (5 min) add: 72 g lime and 48 g $CaSO_4$ or gypsum. Add slowly by hand or it will stick to the sides of the mixer.
   (5) Mix for 20–30 min, stopping mixer two or three times to break up lumps.
   (6) Let mix age for 2 weeks and check pH. Mix pH should be between 5·5 and 6·0. If pH is below 5·2 remix using another 72 g of lime. This should bring pH to above 5·5.

B. *Procedure for making peat–perlite–soil mixture for field planting*
   (1) Use three parts peat–perlite with one part top soil (steam-sterilized).
   (2) Use a good loam soil.
   (3) Mix in cement mixer following the same procedure as outlined for peat–perlite mixture. Add no water or minerals. The pH should be between 5·5 and 6·5.

Adapted from: Feder and Manning.[47]

seventy-two Bel-W3 and seventy-two Bel-B for each major site to be evaluated. The root system should be minimally disturbed when transplanting into the pails or other field settings. Any shock to the plant during transplantation may alter the O_3 response.

The site should be at least 30 m^2 with grid intercepts placed every 15 m. Grid intercepts should be shaded to reduce the light by about 40–50%. A variety of plastic or cotton gauze shade materials are useful for this purpose. Natural shade from trees may be used where appropriate provided that ready access to the wind is ensured. If shade supports are built they should be open on all four sides to ensure multi-directional air movement. Four plants of each cultivar are placed at each intercept for a total of eight plants per intercept and 72 plants per test site. The site should be located with open access to all wind directions. The site should have easy access to water.[47] A self-watering site, patterned after that described by Oshima *et al.*,[106] can be used if water is not readily available. Pails should be buried to the rim in soil to facilitate good soil moisture. Spraying for pests should be carried out on a weekly basis using an insecticide such as malathion, which does not appear to alter plant sensitivity to O_3. Plants should be watered and fertilized weekly with 15–15–15 liquid fertilizer. Wilting should be avoided since water stress alters O_3 sensitivity.[57] After the first four plants have been in place for 3 weeks, two of them are replaced with fresh plants. Thereafter, replace the two oldest plants every three weeks until the end of the growing season. This procedure ensures the availability of newly maturing, O_3-sensitive leaves throughout the growing season.

2. Collection and handling of data

Physical data, such as temperature, rainfall, wind speed and direction, should be recorded. Useful meteorological data for remote areas can sometimes be obtained from the weather service.

Plant injury data should be collected at regular intervals, usually once a week. Each leaf, starting with number No. 1(oldest expanded leaf) should be rated on the same day each week, at about the same time of day and by the same observer. Photographs of degrees of severity of injury, like those prepared by Heck,[56,57] and Heck *et al.*[58] and Oshima *et al.*,[106] can be used to help in estimating percent leaf injury. Characteristic O_3 leaf injury symptoms are shown in Fig. 19.

The observer should visually integrate the injured areas of each leaf and record the percentage of the leaf area injured according to a 0–100% scale, in 5% increments. The total injury score for each date is obtained by adding the individual leaf ratings and entering them in a sum column on a prepared form. Each plant should have a separate record sheet. Dead or completely yellowed leaves are given a rating of 100% injury. The number of new leaves developed between observation periods should also be recorded. New leaf injury is obtained by determining the difference in consecutive readings. A record should be made of (1) the percentage of leaf area injured on each leaf totalled for each plant, (2) new injury on each plant for any given period: record by figuring the difference in cumulative readings between the beginning and end of the period and (3) leaf area of each leaf (length from base to tip times the maximum width of the blade).

The data can be summarized as follows: (1) total injury score—this is the sum of individual leaf ratings for each date—(2) new leaf injury (weekly), (3) new leaf injury (cumulative), (4) total number of leaves after a month, (5) total leaf area after a month. It is relatively simple to interpret the data if it is represented graphically. Several types of curve are possible. Plotting weekly injury scores, or number of leaves injured on a weekly basis against time elapsed gives a cumulative injury curve over time. If an instrument monitor is operational in the vicinity of the study site, cumulative leaf injury can be plotted against cumulative O_3 or against the number of episodes where O_3 levels exceeded 0·04 ppm for more than 4 h (Fig. 20).

Curves for the reactions of Bel-W3 and Bel-B tobacco cultivars exposed to O_3 in a given environment have slopes which can be described mathematically and perceived visually. Comparison of the slopes of these curves describes the quality of the air in terms of O_3 content. Steeply rising slopes for both cultivars indicate the presence of high O_3 concentrations; a steep slope for Bel-W3 and a lower slope for Bel-B indicates a lower O_3 concentration; a still flatter slope for Bel-W3 and a flat slope (no slope) for Bel-B indicates a low O_3 concentration and no slope (flat) for both cultivars means that O_3 is at background level, that is, about 0·02–0·03 ppm for an hourly average. (Note: the new USA Standard for O_3 is 0·12 ppm for one hour, not to be exceeded more than once a year.)

The simultaneous use of sensitive and tolerant cultivar pairs which differ only in their threshold response to O_3 is most important because it allows the generation of two curves whose slopes can then be compared to define air quality.

There are probably other cultivars of tobacco, like SamSun[90,91,95,96] which may prove useful as plant monitors for O_3, but not enough is known

FIG. 21. Range of O_3-induced injury on primary leaves of Tempo bush bean. Injury ranges from none (upper right) to severe (lower left). (Courtesy: W. J. Manning.)

about them to recommend them now. Because of the complications encountered in having to shade Bel-W3 and Bel-B in order to obtain typical plant response, it would be useful to find other cultivars of *Nicotiana* which do not require shading.

B. Bean

A number of cultivars of common garden bean (*Phaseolus vulgaris* L.) are known to be very sensitive to O_3. The older leaves exhibit extensive bronzing and/or stippling followed by chlorosis and leaf fall (Fig. 21). The

TABLE 9

TWO O_3 INJURY RATING SYSTEMS FOR BEANS (*Phaseolus vulgaris* L.) AND EXAMPLES OF THEIR USE

Rating systems *Rating of injury*	*Injury severity index*	*Percentage of leaf injury*
None	0	0
Slight	1	1–25
Moderate	2	26–50
Moderate-severe	3	51–75
Severe	4	76–99
Complete	5	100

Example—primary leaves *Date (days from seed)*	*Injury severity index*	*Percentage of leaf injury*
7	0	0
11	1	15
15	2	40 (+25)
20	4	80 (+40)

Note: Rust infections or mite infestations may result in symptoms on bean leaves that could be confused with O_3 injury.

Injury is rated by visual estimation of percentage of leaf area bronzed or stippled. Cumulative readings can be made at 3–5 day intervals until the leaf falls. (See examples above). Plants should no longer be used after the third set of trifoliate leaves has expanded as it becomes too difficult and time-consuming to keep track of the leaves on a cumulative basis.

Developed by W. J. Manning and adapted from: Feder and Manning.[47]

bush bean Tempo and the dry bean Pinto 111 have been used as O_3 biomonitors. Both are easily grown from seed and do not have special cultural requirements. Under normal garden conditions, beans can be used as short-term biomonitors of O_3. Examples of two O_3 injury systems for beans, and their use, are given in Table 9.

C. Grapes

Some cultivars of fox grape (*Vitis labrusca* L.) can also be used as O_3 biomonitors.[47,77,78,117] Leaf injury or yield can be used as plant responses and the plant is both perennial and woody.

Two commercially available cultivars respond differently to the same O_3 concentration. Cultivar Ives is O_3-sensitive and cultivar Delaware is O_3-tolerant. These two grape cultivars are about as sensitive to O_3 as are two tobacco cultivars, Bel-W3 and Bel-B.

Injury is visually estimated by determining percentage of leaf area affected by small, purplish to black stipples. These areas coalesce to form larger spots as the leaves age (Fig. 22). The leaves eventually become chlorotic, turn brown and may die.[81, 82] The older leaves are the most sensitive and newer leaves are injured in sequence as they mature. The

FIG. 22. Moderate to severe O_3-induced stipple on grape cultivar 'Ives'. (Courtesy: W. J. Manning.)

rating system and method of expressing the data are the same as described for tobacco. However, this is a perennial plant and can, therefore, serve as a monitor at a single site for many years. This means that yield can also be used as a criterion for O_3 response and yields for different years can be compared after factoring out weather and other non-pollutant influences. Unlike tobacco, grapes do not require shading to respond to O_3. One planting can yield many years of data including both the gross yield in terms of grapes harvested per year per hectare and the sugar content of the fruit, factors which are both affected by exposure of the plants to O_3.[47,77,78,119]

D. White Pine

A third type of monitoring plant to be considered would be a tree, like Eastern white pine (*Pinus strobus* L.), an evergreen, coniferous gymnosperm. This species, and Virginia pine (*Pinus virginiana* Mill.), like *N. tabacum* and *V. vinifera*, have strains or cultivars which differ in their response to O_3.[3–8, 22–24] These strains can be selected and multiplied as cuttings or seedlings or grafted plants.[36]

Rootstocks do not influence scion responses to O_3.[36,71–73] This allows for large-scale multiplication in the nursery of clones with the desired O_3 response thresholds. A small nursery planting of these selections can provide large numbers of uniform plants for use in setting out long-term, relatively care-free biomonitors to form a network in areas where it would be impractical to deal with herbaceous and/or other more delicate plant types, and too expensive and impractical to monitor using instrumentation.

Here again, the plant can serve as a monitor by measuring leaf (needle) or twig injury, or by measuring yield in terms of secondary, radial trunk growth. Tip injury to needles, needle mottling, loss of needles and per cent needle retention are all symptoms of O_3 injury.[23,24,36] High levels of O_3 cause severe burning and browning of needle tips. Lower O_3 concentrations cause a chlorotic mottling of recently matured needle tissue, appearing at the tips of the youngest susceptible needles and extending along a greater length of successively older needles.[36,67,68] Late in the season needle drop is stimulated and heavily injured trees may drop all of the older, more mature, needles, leaving only tufts of immature needles at the tips of the twigs. These symptoms are not to be confused with chlorotic dwarf syndrome of Eastern white pine which is apparently caused by a mixture of O_3 and SO_2 and characterized by dwarfed trees with stunted tops and roots, short and twisted mottled needles and premature shedding of foliage[22,70] (see Figs. 23 and 24). CDS, however, is useful for

FIG. 23. Severe incidence of chlorotic dwarf syndrome on Eastern white pine, caused by an $O_3 + SO_2$ interaction, but probably due primarily to O_3. (Courtesy: USDA.)

biomonitoring both O_3 and SO_2 in ambient air to get a picture of air quality.

Sites for monitoring should be selected to ensure: (1) similar soil types and pH, (2) similar drainage conditions, (3) similar elevation and exposure, (4) space to allow full tree growth over many years, (5) easy access to the site. At least eight O_3-susceptible and eight O_3-tolerant trees should be planted at each site. The trees should be placed on 8 m centers in a grid arrangement. Tolerant and susceptible trees should alternate in the grid. The use of the grid allows pooling of data taken from all the trees on a site.

Both growth and injury data should be collected when possible.[46,47,71] Growth data should include: twig length, needle length, needle age, total tree height to terminal tip, and tree diameter or circumference at a designated height above the ground level (e.g. diameter at breast height (dbh)). Injury data should include: needle color (chlorosis, spotting,

FIG. 24. Healthy Eastern white pine. Compare with Fig. 23. (Courtesy: USDA.)

mottling, banding, etc.), needle shape, needle retention and overall tree vigor and appearance.

The long-term performance of a tree can be best measured by comparing trunk diameter and height (growth and growth rate) from year to year. Pollutant stress can also be measured by cone and seed production once the trees reach fruiting age.[67,68] Because of the long-term nature of these measurements and because of the close relationship between temperature and rain and snowfall, and growth and reproduction, it is important to record these parameters.

The comparative rate of twig elongation and trunk enlargement over years between the O_3-tolerant trees, taking edaphic and environmental factors into account, yields a long-term picture of O_3-related air quality. The slopes of curves depicting these growth parameters measured against time reflect the O_3 stress placed upon the test trees. In the case of Ponderosa pine (*Pinus ponderosa* Laws.), east of Los Angeles, California, long-term studies have made possible a correlation between growth rate and actual ambient O_3 concentrations. Whilst this type of correlation cannot yet be made with any degree of certainty in the northeastern USA and in Canada and western Europe, it is, never the less, useful to relate performance of O_3-sensitive and O_3-tolerant Eastern white pine trees to the pollutant stresses present in their air environment. The trees do not quantify the ambient O_3 concentration with fine exactitude, but they do respond to those concentrations in ways that can be measured and are, therefore, meaningful in terms of air quality and pollution abatement.

V. BIOMONITORING PAN

The same general constraints covered under the section on O_3 apply to PAN as far as selecting plant materials, sites, cultural practices, etc. are concerned. PAN has, however, not been as widely recognized, geographically, as being present in ambient air. Much of the research has therefore come from Southern California, where PAN is a regular constituent of photochemical smog and is present in phytotoxic concentrations.[28,37–39,65,80–82,98,109,111,123,125,129,144]

Probably the most useful plants for monitoring PANs are Romaine lettuce (*Lactuca sativa* L.), swiss chard (*Beta chilensis* Hort.) and annual bluegrass (*Poa annua* L.). These herbaceous plants have leaves which are sensitive to PAN but quite tolerant to O_3. Since these two pollutants generally occur together in California it is most useful to have monitors which can respond to O_3 alone or PAN alone. Trees are not PAN-sensitive, which eliminates them as useful monitoring plants.[28]

The earliest symptom of PAN injury is a water-soaking or shininess of the lower leaf surface. As the injury progresses, the spongy mesophyll cells near the stomata collapse and are replaced by air pockets, giving the injured leaf a silvery looking undersurface (Fig. 25). After two or three days, the silvered areas may become bronzed. The other important—and most useful—PAN symptom is referred to as 'banding'. Symptoms appear at the tips of the youngest susceptible leaves (as opposed to the mature leaves injured by O_3). As the leaf tissues grow and mature the injured areas appear as bands. A second exposure to PAN of the same leaf will produce a second set of bands separated from the first band by healthy tissue. Several successive exposures will finally involve the entire leaf.[70,81,82]

It is not clear from the literature whether PAN-tolerant and PAN-sensitive cultivars of lettuce, chard or annual bluegrass have been identified. However, it would not be surprising to find genetic variability to PAN sensitivity amongst cultivars or strains of these three plants. The need exists to identify them in order to be able to set up a monitoring system for characterizing the PAN content of ambient air. This may require breeding and selection to identify useful cultivars or strains. Once this has been done and field studies of the reliability of these sources have been successfully demonstrated it would be no more difficult to set up a PAN plant

FIG. 25. Undersurface leaf glazing on Romaine lettuce leaves, caused by ambient PAN. (Courtesy: USDA.)

monitoring system than it was to develop several for monitoring O_3.

All of the above-mentioned plants can be grown from seed which ensures that large populations could be rapidly produced and maintained once the stocks were selected. After that, the process would be very similar to what was outlined for handling the tobacco/O_3 relationship. Quantification of the data could be best handled by looking at the amount of leaf injury inflicted over time and at the time taken to involve whole leaves on a given plant. Banding would indicate how many episodes of PAN exposure occurred during a given period but banding could be difficult to relate to PAN concentrations in ambient air during an episode. It might be useful to explore how band width relates to PAN concentration.

VI. BIOMONITORING NITROGEN OXIDES

NO_2 and NO play an important role in the formation of O_3 in photochemical smog, but their role as phytotoxicants is not very clear. Apparently only very high concentrations, higher than are usually encountered even in the ambient air of the Los Angeles, California basin, are required to produce plant injury symptoms.[16,17,70,81,82] Summer mean hourly averages in that basin range between 0·08 and 0·15 ppm NO_x, and 0·24 ppm is a mean winter hourly average. If one calculates that less than half of this amount is actually NO_2, then it is difficult to see how any injury would occur on even sensitive plants growing in this 'worst of all cases' area.[98] It takes at least 1·0 ppm for 24 h to produce injury symptoms on leaves of sensitive plants.[81] A study in which sixty species of plants were fumigated with a mixture of 1 : 1 NO and NO_2 showed that NO_2 was only a fifth as toxic as SO_2 and that 6 ppm NO_2 for 4–8 h caused injury to several plant species including peas, bush beans and alfalfa.[81] Mustard (*Brassica arvensis*) is one of the more NO_2-sensitive plant species, but even it would not make a useful monitor in the presence of ambient NO_x concentrations encountered.

VII. COMPARATIVE ADVANTAGES AND DISADVANTAGES OF USING PLANTS TO BIOMONITOR OXIDANTS

A large number of plants have potential as biomonitors of ambient concentrations of O_3. Protocols for monitoring have been well worked out

TABLE 10

COMPARATIVE ADVANTAGES AND DISADVANTAGES OF USING PLANTS TO BIOMONITOR OXIDANTS (PRINCIPALLY O_3)

Advantages
1. Many plant species are sensitive to O_3, and other oxidants.
2. Some cultivars of plants respond to O_3 or PAN in a characteristic fashion.
3. Plant responses can sometimes be related to ambient concentrations of O_3.
4. Some plants have closely related cultivars that differ only in O_3 sensitivity.
5. Biomonitoring with plants is easier to do and cheaper than using expensive instruments in monitoring networks.
Disadvantages
1. Edaphic and environmental factors can affect biomonitoring plant responses to O_3 and PAN.
2. Heavy dependence on symptomatology may lead to errors in determining injury on biomonitor plants.
3. Greater quantification of plant responses to O_3 or PAN is needed to draw meaningful conclusions about dose/response relationships.
4. Better biomonitors for PAN are needed.
5. Instrument monitors cannot be eliminated entirely. It is necessary to know ambient concentrations of O_3 and PAN.

for several useful plants. A few plants seem to be useful in biomonitoring PAN, whilst, at the moment, there does not seem to be a need to biomonitor NO_x. A summary of the comparative advantages and disadvantages of using plants to biomonitor oxidants (particularly O_3) is presented in Table 10.

REFERENCES

1. BENEDICT, H. M. and W. H. BREEN. (1955). The use of weeds as a means of evaluating vegetation damage caused by air pollution. *Proc. Nat. Air Poll Symp.* (3rd) pp. 117–90.
2. BENNET, J. P. and R. J. OSHIMA. (1976). Carrot injury and yield response to ozone. *J. Am. Soc. Hort. Sci.*, **101**: 638–9.
3. BERRY, C. R. (1964). Eastern white pine: A tool to detect air pollution. *Southern Lumberman* (December).
4. BERRY, C. R. (1971). Relative sensitivity of red, jack, and white pine seedlings to ozone and sulfur dioxide. *Phytopathology*, **61**: 231–2.
5. BERRY, C. R. (1973). The differential sensitivity of Eastern white pine to three types of air pollution. *Can. J. Forest Rec.*, **3**: 543–7.
6. BERRY, C. R. (1974). Age of pine seedlings with primary needles affects sensitivity to ozone and sulfur dioxide. *Phytopathology*, **64**: 207–9.
7. BERRY, C. R., and H. E. HEGGESTAD. (1968). Air Pollution Detectives. *Yearbook of Agriculture*, USDA, pp. 142–6.
8. BERRY, C. R. and L. A. RIPPERTON. (1963). Ozone, a possible cause of white pine emergence tipburn. *Phytopathology*, **53**: 552–7.
9. BRASHER, E. P., D. J. FIELDHOUSE and M. SASSER. (1973). Ozone injury in potato variety trails. *Plant Disease Reptr.*, **57**: 542–4.
10. BRENNAN, E. and P. M. HALISKY. (1970). Response of turfgrass cultivars to ozone and sulfur dioxide in the atmosphere. *Phytopathology*, **60**: 1544–6.
11. BRENNAN, E., I. A. LEONE and R. H. DAINES. (1964). The importance of variety in ozone plant damage. *Plant Disease Reptr.*, **48**: 923–4.
12. CAMERON, J. W. (1975). (Personal Communication).
13. CAMERON, J. W. (1975). Inheritance in sweet corn for resistance to acute ozone injury. *J. Am. Soc. Hort. Sci.*, **100**: 577–9.
14. CAMERON, J. W., H. JOHNSON, O. C. TAYLOR and H. W. OTTO. (1970). Differential susceptibility of sweet corn hybrids to field injury by air pollution. *HortScience*, **5**: 217–19.
15. CAMERON, J. W. and O. C. TAYLOR. (1973). Injury to sweetcorn inbreds and hybrids by air pollutants in the field and by ozone treatments in the greenhouse. *J. Environ. Qual.*, **2**: 387–9.
16. CAPRON, T. M. and T. A. MANSFIELD. (1975). Generation of nitrogen oxide pollutants during CO_2 enrichment of glasshouse atmospheres. *J. Hort. Sci.*, **50**: 233–8.
17. CAPRON, T. M. and T. A. MANSFIELD. (1976). Inhibition of net photosynthesis in tomato in air polluted with NO and NO_2. *J. Exp. Bot.*, **27**:1181–6.

18. Clayberg, C. D. (1971). Screening tomatoes for ozone resistance. *HortScience*, **6**: 396–7.
19. Clayberg, C. D. (1972). Evaluation of tomato varieties for resistance to ozone. *Conn. Agr. Exp. Sta. Circular* 246, 11pp.
20. Cleveland, W. S. and B. Kleiner. (1975). transport of photochemical air pollution from Camden–Philadelphia urban complex. *Environ. Sci. and Tech.*, **9**: 869–72.
21. Cleveland, W. S., B. Kleiner, J. E. McRae and J. L. Warner. (1976). Photochemical air pollution: Transport from New York City area into Connecticut and Massachusetts. *Science*, **191**: 179–81.
22. Costonis, A. C. (1973). Injury to Eastern White Pine by sulfur dioxide and ozone alone and in mixtures. *Eur J. Forest Pathol.*, **3**: 50–55.
23. Costonis, A. C. and W. A. Sinclair. (1969). Relationship of atmospheric ozone to needle blight of Eastern white pine. *Phytopathology*, **59**: 1566–74.
24. Costonis, A. C. and W. A. Sinclair. (1969). Ozone injury to *Pinus strobus*. *J. Air Poll. Contr. Assoc.*, **19**: 867.
25. Craker, L. E. and W. A. Feder. (1972). Development of the inflorescence in petunia, geranium, and poinsettia under ozone stress. *HortScience*, **7**: 59–62.
26. Daines, R. H., I. A. Leone, E. Brennan and J. T. Middleton. (1960). Damage to spinach and other vegetation in New Jersey from ozone and other air-borne oxidants. *Phytopathology.*, **50**: 570.
27. Darley, E. F. (1960). Use of plants for air pollution monitoring. *J. Air Poll. Contr. Assoc.*, **10**: 198–9.
28. Davis, D. D. (1975). Resistance of young Ponderosa pine seedlings to acute doses of PAN. *Plant Disease Reptr.*, **59**: 183–4.
29. Davis, D. D. and J. B. Coppolino. (1974). Relative ozone susceptibility of selected woody ornamentals. *HortScience*, **9**: 537–9.
30. Davis, D. D. and J. B. Coppolino. (1976). Ozone susceptibility of selected woody shrubs and vines. *Plant Disease Reptr.*, **60**: 876–8.
31. Davis, D. D. and H. D. Gerhold. (1976). Selection of trees for tolerance of air pollutants. In: *Better Trees for Metropolitan Landscapes*. USDA Forest Service, Gen Tech. Report No. 22 pp. 61–66.
32. Davis, D. D. and L. Kress. (1974). The relative susceptibility of ten bean varieties to ozone. *Plant Disease Reptr.*, **58**: 14–16.
33. Davis, D. D. and F. A. Wood. (1972). The relative susceptibility of eighteen coniferous species to ozone. *Phytopathology*, **62**: 14–19.
34. Davis, D. D. and R. G. Wilhour. (1976). *Susceptibility of woody plants to sulfur dioxide and photochemical oxidants*. US EPA Publication 600/3–76–102.
35. Demeritt, M. E., W. M. Chang, J. D. Murphy and H. D. Gerhold. (1971). Selection system for evaluating resistance of Scotch pine seedlings to ozone and sulfur dioxide. *Proc. NE Forest Tree Improvement Conf., Univ. Maine, Orono*, pp. 87–97.
36. Dochinger, L. S. (1968). The impact of air pollution on Eastern white pine: The chlorotic dwarf disease. *J. Air Poll. Contr. Assoc.*, **18**: 814–16.
37. Dugger, W. M., O. C. Taylor, W. H. Klein and W. Shropshire. (1963). Action spectrum of peroxyacetyl nitrate damage to bean plants. *Nature (London)*, **198**: 75–76.

38. DUGGER, W. M., O. C. TAYLOR, C. R. THONPSON and E. CARDIFF. (1963). The effect of light on predisposing plants to ozone and PAN damage. *J. Air Poll. Contr. Assoc.*, **13**: 423–8.
39. DUGGER, W. M. and I. P. TING. (1968) The effect of peroxyacetyl nitrate on plants: Photoreductive reactions and susceptibility of bean plants to PAN. *Phytopathology*, **58**: 1102–7.
40. DUNNING, J. A. and W. W. HECK. (1973). Response of Pinto bean and tobacco to ozone as conditioned by light intensity and/or humidity. *Environ. Sci. and Tech.* **7**: 824–6.
41. ENGLE, R. L. and W. H. GABELMAN. (1966). Inheritance and mechanism for resistance to ozone damage in onion, *Allium cepa* L. *Proc. Am. Soc. Hort. Sci.*, **89**: 423–30.
42. FEDER, W. A. (1970). *Chronic effects of low levels of air pollutants upon floricultural and vegetable plants in the northeast.* Ann. Rept. US Public Health Service Contract No. PH 22–68–39.
43. FEDER, W. A. (1970). Plant response to chronic exposure to low levels of oxidant type air pollution, *Environ. Pollut.* **1**: 73–76.
44. FEDER, W. A. (1970). Modifying the environment. *HortScience*. **5**: 247–9.
45. FEDER, W. A. (1977). Adverse effects of chronic low level ozone exposure to tomato fruit set and yield. In:. *Proc. 100th Anniv. Cottrell Symp. Calif. State Univ., Stanislaus* (Perona, M. (ed.)), pp. 153–155.
46. FEDER, W. A. (1978). Plants as bioassay systems for monitoring atmospheric pollutants. *Environ. Health Perspectives*, **27**: 139–47.
47. FEDER, W. A. and W. J. MANNING (1979). Living plants as indicators and monitors. In:, *Handbook of Methodology for the Assessment of Air Pollution Effects on Vegetation.* (Heck, W. W., Krupa, S. V. and S. N. Linzon (eds.)), Air Poll. Contr. Assoc., Pittsburgh, Pa., pp. 9–1 to 9–14.
48. FEDER, W. A. and F. J. CAMPBELL. (1968). Influence of low levels of ozone on flowering of carnations. *Phytopathology*, **58**: 1038.
49. FEDER, W. A., F. L. FOX, W. W. HECK and F. J. CAMPBELL. (1969). Varietal responses of petunia to several air pollutants. *Plant Disease Reptr.* **53**: 506–10.
50. FEDER, W. A., T. J. KELLEHER, W. D. RILEY and I. PERKINS. (1975). Ozone injury on tobacco plants on Nantucket Island is caused by long-range transport of ozone from the mainland. *Proc. Am. Phytopathol.* Soc., **2**: 97.
51. FLOOR, H. and A. C. POSTHUMUS. (1977). Biologische erfassung von ozone und PAN-Immissionen In den Niederland 1973, 1974, and 1975. *VDI-Berichte*, **270**: 183–90.
52. FUGIWARA, T. (1973). Effects of nitrogen oxides in the atmosphere on vegetation. *J. Pollut. Contr.*, **9**: 253–7.
53. GENTILE, A. G., W. A. FEDER, R. E. YOUNG and Z. SANTER. (1971). Susceptibility of *Lycopersicon* spp. to ozone injury. *J. Am. Soc. Hort. Sci.*, **96**: 94–96.
54. HEAGLE, A. S., D. E. BODY and E. K. POUNDS. (1972). Effect of ozone on yield of sweet corn. *Phytopathology*, **62**: 683–7.
55. HEAGLE, A. S., D. E. BODY and G. E. NEELY. (1974). Injury and yield responses of soybean to chronic doses of ozone and sulfur dioxide in the field. *Phytopathology*, **64**: 132–6.

56. HECK, W. W. (1966). The use of plants as indicators of air pollution. *Air and Water Pollut.*, **10**: 99–111.
57. HECK, W. W. (1968). Factors influencing expression of oxidant damage to plants. *Ann. Rev. Phytopathol.*, **6**: 165–88.
58. HECK, W. W., J. A. DUNNING and I. J. HINDAWI (1966). Ozone: Non-linear relation of dose and injury in plants. *Science*, **151**: 577–8.
59. HECK, W. W., F. L. FOX, C. S. BRANDT and H. A. DUNNING. (1969). *Tobacco, a sensitive monitor for photochemical air pollution.* US. Nat. Air Poll. Contr. Admin. Publication AP-55.
60. HECK, W. W. and A. S. HEAGLE. (1970). Measurement of photochemical air pollution with a sensitive monitoring plant. *J. Air Poll. Contr. Assoc.*, **20**: 97–99.
61. HECK, W. W., and D. T. TINGEY. (1970). Time-concentration model to predict acute foliar injury. *Proc. Int. Clean Air Congress, Washington, DC. Ind*: 249–255.
62. HEGGESTAD, H. E. (1973). Photochemical air pollution injury to potatoes in the Atlantic coastal states. *Am. Potato J.*, **50**: 315–28.
63. HEGGESTAD, H. E. and W. W. HECK. (1971). Nature, extent, and variation of plant response to air pollutants. *Adv. Agron.*, **23**: 111–45.
64. HEGGESTAD, H. E. and H. A. MENSER. (1962). Leaf spot-sensitive tobacco strain Bel-W3, a biological indicator of the air pollutant ozone. *Phytopathology*, **52**: 735.
65. HEGGESTAD, H. E., and E. F. DARLEY. (1968). Plants as indicators of the air pollutants ozone and PAN. *Proc. First Eur. Congr. Influence of Air Poll. on Plants, and Animals, Wageningen, Netherlands*, pp. 329–40.
66. HINDAWI, I. J. (1970). *Air Pollution Injury to Vegetation.* US., Nat Air Poll. Contr. Assoc. Publication No. AP-71.
67. HOUSTON, D. B. (1974). Response of selected *Pinus strobus* L. clones to fumigations with sulfur dioxide and ozone, *Can. J. For. Res.*, **4**: 65–68.
68. HOUSTON, D. B. and G. R. STAIRS. (1973). Genetic control of sulfur dioxide and ozone tolerance in Eastern white pine. *Forest Sci.*, **19**: 267–271.
69. HOWELL, R. K., T. E. DEVINE and C. H. HANSON. (1971). Resistance of selected alfalfa strains to ozone. *Crop Sci.*, **11**: 114–5.
70. JACOBSON. J. S. and A. C. HILL. (1970). *Recognition of Air Pollution Injury to Vegetation: A Pictorial Atlas.* Air Poll. Contr. Assoc., Pittsburgh, Pa.
71. JENSEN, K. F. (1973). Response of nine forest tree species to chronic ozone fumigation. *Plant Disease Reptr.* **57**: 914–17.
72. JENSEN, K. F. and L. S. DOCHINGER. (1974). Responses of hybrid poplar cuttings to chronic and acute levels of ozone. *Environ. Pollut.*, **8**: 289–95.
73. JENSEN, K. F. and R. G. MASTERS. (1975). Growth of six woody species fumigated with ozone. *Plant Disease Reptr.*, **59**: 760–2.
74. JOHNSON, H., J. W. CAMERON and O. C. TAYLOR. (1971). Air pollution resistance in sweet corn varieties. *Calif. Agric.*, **25**: 8–10.
75. JUHREN, M., W. M. NOBLE and F. W. WENT. (1957). The standardization of *Poa annua* as an indicator of smog concentrations. 1. Effects of temperature, photoperiod, and light intensity during growth of test plants. *Plant Physiol.*, **32**: 576–88.

76. KELLEHER, T. J. and W. A. FEDER. (1978). Phytotoxic concentrations of ozone on Nantucket Island: Long range transport from the Middle Atlantic States over the open ocean confirmed by bioassay with ozone sensitive tobacco plants. *Environ. Pollut.*, **17**: 187–94.
77. KENDER, W. J. and S. C. CARPENTER. (1974). Susceptibility of grape cultivars and selections to oxidant injury, *Fruit Var. J.*, **28**: 59–61.
78. KENDER, W. J. and N. J. SHAULIS. (1976). Vineyard management practices influencing oxidant injury in 'Concord' grapevines. *J. Am. Soc. Hort. Sci.*, **101**: 129–32.
79. KNABE, W., C. S. BRANDT, H. VAN HAUT and C. J. BRANDT. (1973). Nachweis photochemischer Luftverunreinigungen durch biologische Indikatoren in der Bundesrepublik Deutschland. *Proc. 3rd Int. Clean Air Congress, Dusseldorf.* pp. A110–A114.
80. KOHUT, R. J. (1972). *Response of hybrid poplar to simultaneous exposure to ozone and PAN.* Center for Air Environ. Studies Publication No. 288–72. Penn. State Univ., Univ. Park, Pa. 26 p.
81. LACASSE, N. L. and W. J. MOROZ (eds.). (1969). *Handbook of effects assessment vegetation damage.* Center for Air Environ. Studies, Penn. State Univ., Univ. Park, Pa.
82. LACASSE, N. L. and M. TRESHOW. (1976). *Diagnosing vegetation injury caused by air pollution.* US EPA Publication. 140 pp.
83. LINZON, S. N. (1966). Damage to eastern white pine by sulfur dioxide, semimature tissue needle blight and ozone. *J. Air Poll. Contr. Assoc.* **16**: 140–5.
84. MACLEAN, D. C. and R. E. SCHNEIDER. (1976). Photochemical oxidants in Yonkers, New York: Effects on yield of bean and tomato. *J. Environ. Qual.*, **5**: 75–78.
85. MACDOWALL, F. D. H., E. I. MUKAMMAL and A. F. W. COLE. (1964). Direct correlation of air-polluting ozone and tobacco weather fleck. *Can. J. Plant Sci.*, **44**: 410–17.
86. MACDOWALL, F. D. H. (1965). Predisposition of tobacco to ozone damage *Can J. Plant Sci.*, **45**: 1–12.
87. MANNING, W. J. and W. A. FEDER. (1976). Effects of ozone on economic plants. In:. *Effects of Air Pollutants on Plants.* (Mansfield, T. A. (ed)), Cambridge Univ. Press, Cambridge,
88. MANNING, W. J. and J. F. MCCARTHY. (1976). Indicator plants for detection of atmospheric oxidants in Massachusetts: Alternatives to Bel-W3 tobacco. *Proc. Am. Phytopathol. Soc.*, **3**: 308.
89. MANNING, W. J., W. A. FEDER and I. PERKINS. (1972). Sensitivity of spinach cultivars to ozone. *Plant Disease Reptr.* **56**: 832–33.
90. MENSER, H. A. (1966). Response to ozone of five flue-cured tobacco varieties. *Tobacco Sci.*, **10**: 33–34.
91. MENSER, H. A. (1969). Effects of air pollution on tobacco cultivars grown in several states. *Tobacco*, **169**: 20–25.
92. MENSER, H. A., J. J. GRASSO, H. E. HEGGESTAD and O. E. STREET. (1964). Air filtration study of "hidden" air pollution injury to tobacco plants. *Plant Physiol.*, **39**: 18.

93. MENSER, H. A., H. E. HEGGESTAD, O. E. STREET and R. N. JEFFREY. (1963). Response of plants to air pollutants. I: Effects of ozone on tobacco plants preconditioned by light and temperature. *Plant Physiol.*, **38**: 605–9.
94. MENSER, H. A., H. E. HEGGESTAD and O. E. STREET. (1963). Response of plants to air pollutants. II. Effects of ozone concentration and leaf maturity on injury to *Nicotiana tabacum*. *Phytopathology*, **53**: 1304–9.
95. MENSER, H. A. and G. H. HODGES. (1968). Varietal tolerance of tobacco to ozone dose rate. *Agron. J.*, **60**: 349–52.
96. MENSER, H. A., G. H. HODGES and C. G. MCKEE. (1973). Effects of air pollution on Maryland (Type 32) tobacco. *J. Environ. Qual.*, **2**: 253–8.
97. MILLER, P. R., J. R. PARMETER, O. C. TAYLOR and E. A. CARDIFF. (1963). Ozone injury to the foliage of *Pinus ponderosa*. *Phytopathology*, **53**:1072–6.
98. MUDD, J. B. (1975). Peroxyacetyl nitrates. In: *Responses of Plants to Air Pollution* (Mudd, J. B. and Press. T. T. (eds.)), Academic Press, pp. 97–119.
99. MURRAY, J. J., R. K. HOWELL and A. C. WILTON. (1975). Differential response of seventeen *Poa pratensis* cultivars to ozone and sulfur dioxide. *Plant Disease Reptr.*, **59**: 852–4.
100. ORMROD, D. P. (1976). Sensitivity of pea cultivars to ozone. *Plant Disease Reptr.*, **60**: 423–6.
101. ORMROD, D. P. and N. O. ADEDIPE. (1975). Experimental exposures and crop monitors to confirm air pollution. *HortScience*, **10**: 493–4.
102. ORMROD, D. P., N. O. ADEDIPE and G. HOFSTRA. (1971). Responses of cucumber, onion and potato cultivars to ozone. *Can J. Plant Sci.* **51**: 283–8.
103. OSHIMA, R. J. (1974). A viable system of biological indicators for monitoring air pollutants. *J. Air Poll. Contr. Assoc.*, **24**: 576–8.
104. OSHIMA, R. J. (1976). Ozone dosage-crop loss function for alfalfa: A standardized method for assessing crop losses from air pollutants. *J. Air Poll. Contr. Assoc*, **26**: 861–5.
105. OSHIMA, R. J. (1978). (Personal Communication).
106. OSHIMA, R. J., O. C. TAYLOR, P. K. BRAEGELMANN and D. W. BALDWIN. (1975). Effect of ozone on the yield and plant biomass of a commercial variety of tomato. *J. Environ. Qual.*, **4**: 483–4.
107. OSHIMA, R. J., P. K. BRAEGELMANN, D. W. BALDWIN, V. VANWAY and O. C. TAYLOR. (1977). Responses of five cultivars of fresh market tomato to ozone: A contrast of cultivar screening and foliar injury and yield. *J. Am. Soc. Hort. Sci.*, **102**: 286–9.
108. OSHIMA, R. J., P. K. BRAEGELMANN, D. W. BALDWIN, V. VANWAY and O. C. TAYLOR (1977). Reduction of tomato fruit size and yield by ozone. *J. Am. Soc. Hort. Sci.*, **102**: 289–93.
109. PEARSON, R. G., D. B. DRUMMOND, W. D. MCILVEEN and S. N. LINZON. (1974). PAN-type injury to tomato crops in southwestern Ontario. *Plant Disease Reptr.* **58**: 1105–8.
110. POSTHUMUS, A. C. (1976). The use of higher plants as indicators of air pollution in the Netherlands. In: *Proc. Kuopio Mtg. Plant Damages Caused by Air Pollut.* (Karenlampl, L. (ed.)), Kuopio, Finland.
111. POSTHUMUS, A. C. (1977). Experimentelle untersuchungen der wirkung von ozon und peroxyacetylnitrat (PAN). *VDI-Berichte, No.* 270., pp. 153–61.
112. PRASAD, K., J. L. WEIGLE and C. M. SHERWOOD. (1970). Variation in ozone

sensitivity among *Phaseolus vulgaris* cultivars. *Plant Disease Reptr.* **54**: 1026–9.

113. RAJPUT, C. B. S. and D. P. ORMROD. (1976). Response of eggplant cultivars to ozone. *HortScience*, **11**: 462–3.
114. REINERT, R. A. (1975) Monitoring, detecting and effects of air pollutants on horticultural crops: sensitivity of genera and species. *HortScience* **10**: 495–500.
115. REINERT, R. A., D. T. TINGEY, and H. B. CARTER. (1972). Sensitivity of tomato cultivars to ozone. *J. Am. Soc. Hort. Sci.*, **97**: 149–51.
116. RICH, S. (1964). Ozone damage to plants. *Ann. Rev. Phytopathol.*, **2**: 253–66.
117. RICHARDS, B. L., J. T. MIDDLETON and W. B. HEWITT. (1958). Air pollution with relation to agronomic crops. V. Oxidant stipple of grape. *Agron. J.*, **50**: 559–61.
118. ROBERTS, B. R. (1976). The response of field-grown white pine seedlings to different sulphur dioxide environments. *Environ. Pollut.*, **11**: 175.
119. SHAULIS, N. J., W. J. KENDER, C. PRATT and W. A. SINCLAIR. (1972). Evidence for injury by ozone in New York vineyards. *HortScience*, **7**: 570–2.
120. SPIERINGS, F. H. F. G. (1971). Influence of fumigations with NO_2 on growth and yield of tomato plants. *Neth. J. Plant Pathol.*, **77**: 194–200.
121. STARKEY, T. E. (1975). *The influence of peroxyacetyl nitrate on bean* (Phaseolus vulgaris L.) *subjected to post-exposure water stress.* Center for Air Environ. Studies Publication 400–75, Penn. State Univ., Univ. Park, Pa. 45 p.
122. STARKEY, T. E., D. D. DAVIS and W. MERRILL. (1976). Symptomatology and susceptibility of ten bean varieties exposed to peroxyacetyl nitrate (PAN). *Plant Disease Reptr.*, **60**: 480–3.
123. TAYLOR, O. C. (1969). Importance of peroxyacetyl nitrate (PAN) as a phytotoxic air pollutant. *J. Air Poll. Contr. Assoc.*, **19**: 347–51.
124. TAYLOR, O. C. and F. M. EATON. (1966). Suppression of plant growth by nitrogen dioxide. *Plant Physiol.*, **41**: 132–5.
125. TAYLOR, O. C. and D. C. MACLEAN. (1970). Nitrogen oxides and the peroxyacetyl nitrates In:. *Recognition of Air Pollution Injury to Vegetation.* (Jacobson, J. S. and Hill, A. C. (eds.)), Air Poll. Contr. Assoc., Pittsburgh, Pa. pp. E 1 to E 14.
126. TAYLOR, O. C., E. R. STEPHENS, E. F. DARLEY and E. A. CARDIFF. (1960). Effect of air-borne oxidants on leaves of pinto bean and petunia. *Proc. Am. Soc. Hort. Sci.*, **75**: 435–44.
127. TAYLOR, G. S. (1974). Ozone injury on tobacco seedlings can predict susceptibility in the field. *Phytopathology*, **64**: 1047–8.
128. THOMAS, M. D. (1958). Air pollution with relation to agronomic crops. I. General states of research on the effects of air pollution on plants. *Agron. J.*, **50**: 545–50.
129. THONSON, W. W., W. M. DUGGER and R. L. PALMER. (1965). Effects of peroxyacetyl nitrate on ultrastructure of chloroplasts. *Bot. Gaz.*, **126**: 66–72.
130. THOMPSON, C. R., E. HENSEL and G. KATS. (1969). Effects of photochemical air pollutants on Zinfandel grapes. *HortScience*, **4**: 222–4.
131. THOMPSON, C. R. and G. KATS. (1970). Antioxidants reduce grape yield reductions from photochemical smog. *Calif. Agric.*, **24**: 12–13.
132. THOMPSON, C. R., E. G. HENSEL, G. KATS and O. C. TAYLOR. (1970). Effects

of continuous exposure of navel oranges to NO_2. *Atmos. Environ.*, **4**: 349–55.

133. THOMPSON, C. R., G. KATS and E. G. HENSEL. (1971). Effects of ambient levels of NO_2 on navel oranges. *Environ. Sci. & Technol.*, **5**: 1017–19.
134. THOMPSON, C. R. and G. KATS. (1975). Effects of ambient concentrations of peroxyacetyl nitrate on navel orange trees. *Environ. Sci. & Technol.*, **9**: 35–38.
135. THOMPSON, C. R., G. KATS and J. W. CAMERON. (1976). Effects of ambient photochemical air pollutants on growth, yield, and ear characteristics of two sweet corn hybrids. *J. Environ. Qual.*, **5**: 410–12.
136. TINGEY, D. T. (1979). (Personal Communication).
137. TINGEY, D. T., R. C. FITES and C. WICKLIFT. (1973). Foliar sensitivity of soybeans to ozone as related to several leaf parameters. *Environ. Pollut.*, **4**: 183–92.
138. TOWNSEND, A. M. (1974). Sorption of ozone by nine shade tree species. *J. Am Soc. Hort. Sci.*, **99**: 206–8.
139. WALKER, J. T. and J. C. BARLOW. (1974). Response of indicator plants to ozone levels in Georgia. *Phytopathology*, **64**: 1122–7.
140. WEAVER, G. M. and H. O. JACKSON. (1968). Relationship between bronzing in white beans and phytotoxic levels of atmospheric ozone in Ontario, *Can. J. Plant Sci.*, **48**: 561–8.
141. WILHOUR, R. G. (1971). *The influence of ozone on white ash.* Center for Air Environ. Studies, Publication No. 188–71, Penn. State Univ., Univ. Park, Pa.
142. WILTON, A. C., J. J. MURRAY, H. E. HEGGESTAD and F. V. JUSKA. (1972). Tolerance and susceptibility of Kentucky bluegrass (*Poa pratensis* L.) cultivars to air pollution: in the field and in an ozone chamber. *J. Environ. Qual.*, **1**: 112–14.
143. WOOD, F. A., and J. B. COPPOLINO. (1972). The influence of ozone on deciduous tree species. *Mitti Forstl. Bundesversuchanst., Wien.* **97**: 233–53.
144. WOOD, F. A. and D. B. DRUMMOND. (1974). Response of eight cultivars of chrysanthemum to peroxyacetyl nitrate. *Phytopathology*, **64**: 897–8.
145. VANHAUT, H. and H. STRATTMAN. (1974). Experimental investigation of the effect of nitrogen dioxide on plants. *Trans. Land Inst. Pollut. Contr. Soil Conserv. Land North Rhine-Westphalia, Essen* **7**: 50–70.

CHAPTER 5

Biomonitoring Sulfur Dioxide

I. INTRODUCTION

Sulfur dioxide (SO_2) is a well known and extensively studied air pollutant. Its release during power generation, manufacturing, refining and smelting has resulted in many elaborate and thorough field studies on acute SO_2 injury on vegetation near point sources and chronic injury on plants farther away. By examining vegetation near an SO_2 source, a map showing plant injury can be constructed.[38] The shape of the affected area on the map reflects the extent of the SO_2 source and the influence of prevailing winds and topography on SO_2 dispersion. When the SO_2 source is isolated, the mapped area may be ellipsoidal and may cover only a few square miles. Multiple point sources, however, may result in a map that covers several hundred square miles.

Native vegetation is often used to determine the presence of SO_2.[6] A number of trees, shrubs and herbs are known to respond to SO_2 in predictable ways. Lower plants, such as lichens, mosses and fungi, are also sensitive to SO_2 and some make excellent indicators or biomonitors of atmospheric SO_2. Both higher vascular plants and lower non-vascular plants will be considered here.

II. DIAGNOSING INJURY ON HIGHER PLANTS

Many plants are known to be injured by SO_2 under natural and experimental conditions. A number of lists and Tables have been published in which plants are ranked according to their relative sensitivity to SO_2.[2,7,22,38,40,48] Descriptions of acute and chronic SO_2 symptoms are given in these publications as well as in Chapter 2 of this volume. Acute SO_2 injury is also illustrated in Figs 26 to 29.

FIG. 26. Acute injury on white ash leaves (*Fraxinus americana* L.), (left), and a white birch leaf (*Betula pendula* Roth.) (right), caused by sulfur dioxide (SO_2). (Courtesy: USDA and US EPA.)

FIG. 27. Acute injury on dewberry leaves (*Rubus* sp.), caused by sulfur dioxide (SO_2). (Courtesy: US Forest Service.)

FIG. 28. Acute injury on trembling aspen leaves (*Populus tremuloides* Michx.), caused by sulfur dioxide (SO_2). Injury is both marginal and interveinal. (Courtesy USDA.)

FIG. 29. Burned tips on current season's needles of a pine (*Pinus* sp.), caused by sulfur dioxide (SO_2). (Courtesy: US Forest Service.)

TABLE 11

SENSITIVE BIOINDICATORS OF ATMOSPHERIC SULFUR DIOXIDE IN RELATION TO THE TIME OF THE GROWING SEASON WHEN THEY ARE LIKELY TO BE THE MOST SENSITIVE—AND HENCE THE MOST USEFUL

Spring and early summer
Annual bluegrass (*Poa annua* L.)
Mustard (*Brassica* spp.)
Violet (*Viola* spp.)
Zinnia (*Zinnia elegans* Jacq.)
Bracken fern (*Pteridium* spp.)
Blackberry and raspberry (*Rubus* spp.)
Foxgrape (*Vitis vulpina* L.)
Crabapple and apple (*Malus* spp.)
Trembling aspen (*Populus tremuloides* Michx.)
White ash (*Fraxinus americana* L.)
White birch (*Betula pendula* Roth.)
Summer
Alfalfa (*Medicago sativa* L.)
Barley (*Hordeum vulgare* L.)
Buckwheat (*Fagopyrum esculentum* Moench.)
Endive (*Cichorium endiva* L.)
Pumpkin (*Cucurbita pepo* L.)
Squashes (*Cucurbita* spp.)
Late summer
Eastern white pine (*Pinus strobus* L.)
Jack pine (*Pinus banksiana* Lamb.)
Norway spruce (*Picea abies* L.)

Adapted from: references 2, 14, 38 and 52

It is essential to have a good working knowledge of which plant species and varieties are injured by SO_2 and when during the growing season they are most sensitive to SO_2 under field conditions. Table 11 gives some common bioindicators used in field studies. On some indicator plants, acute SO_2 injury may be very typical and predictable. Injury on other plants, especially chronic injury, may not be so obvious.

A. Symptoms

Characteristic foliar markings on leaves of plants known to be sensitive to SO_2 are often used as evidence of atmospheric SO_2. There are, however, a number of factors which must be considered when using this method.[2, 37] Things to look for or determine include:

(1) Localized source(s) of SO_2

(2) What kinds of plants are involved?

Are they known to be sensitive to SO_2?

(3) What kinds of symptoms are observed?
Are they characteristic of SO_2 injury?

(4) Are symptoms more severe near the suspected source?
Do they decrease with distance from the source?

(5) Is there a relationship between incidences and severity of symptoms on plants and prevailing wind patterns?

Incidence and severity of symptoms on known SO_2-sensitive plants are sometimes used in conjunction with instrument monitoring stations to demonstrate SO_2 concentrations at the station and how this relates to vegetation injury. Linzon[39] was able to correlate data from Thomas autometer air sampling stations and SO_2 injury on Eastern white pine (*Pinus strobus* L.) in the Sudbury, Ontario, Canada, area. Based on tree injury severity and air sampling data, he was able to establish three zones of study—inner, intermediate and outer—in relation to distance from SO_2 sources in Sudbury.

B. Chemical Analyses

Unlike most other air pollutants, higher plants have a requirement for sulfur. Most plant sulfur comes from soil, but herbaceous and woody plants also accumulate sulfur from the air via foliar absorption of SO_2 from the atmosphere.[27,41,45,46] This removal of SO_2 by plant leaves is affected by wind velocity, light intensity and plant height or canopy.[27] Sulfur accumulates in plant tissues and standard chemical analyses can be used to determine total sulfur or sulfate in plant tissues. Van Raay[52] determined the sulfur content of herbaceous plants from biomonitoring plots in Holland and expressed it as per cent sulfur in total dry matter. Lihnell[37] mapped SO_2 in Sweden by quantifying sulfate in birch leaves as determined by turbidometric measurements of the density of precipitated barium sulfate from leaf samples.

Limitations to the use of sulfur concentrations in plant leaves as an indication of SO_2 injury and severity have been summarized by Linzon *et al.*[41]:

(1) Sulfur is a normal plant constituent. Background levels must be known. These change throughout the season and at different locations, depending on plant species, site factors and stage of plant growth.

(2) The normal sulfur concentration of the soil must be known.

(3) The location of the SO_2 source must be known.

(4) Air monitoring data for SO_2 should be available.

Under ideal conditions, information on symptom expression on sensitive plants would be combined with chemical analyses for sulfur or sulfate in plant parts and correlated to air monitoring data for SO_2 concentrations. This would allow for accurate diagnosis of SO_2 injury to vegetation and quantitative determination of the effects of SO_2 on vegetation in a given area.

III. BIOMONITORING SO_2 WITH HIGHER PLANTS

A. Eastern White Pine

1. Chlorotic dwarf syndrome

Under nursery and natural conditions, a certain percentage of Eastern white pine seedlings develop a striking disease called the chlorotic dwarf syndrome (CDS). This disease is characterized by a light green color of the new needles which then spot and mottle and yellow and later become stunted and twisted with burned tips. Older needles fall before new ones mature. Trees are severely stunted and usually die. CDS is caused by a combination of SO_2 and O_3.[9–11,28,46] SO_2 alone may cause needle tips to turn red, then brown and fade to grey. Isolated yellow spots or mottling in needles may also occur. (Fig. 29).

Dochinger *et al.*[11] feel that SO_2 is the more reactive of the two gases in causing CDS symptoms. Roberts[46] compared CDS-sensitive and CDS-tolerant white pine clones in the field in high and low SO_2 areas. Using iodometric sulfur determinations of leaf, stem and root tissues, he was able to demonstrate that tolerant trees had higher sulfur contents in the high SO_2 area than in the low SO_2 area. No significant differences in total sulfur content for susceptible trees in either the high or low SO_2 areas were observed.

2. Long-term volume growth loss

Linzon[39] considers Eastern white pine to be the most SO_2-sensitive conifer. Repeated SO_2 injury over time weakens trees and results in tree death.

Using data on the severity of SO_2 injury on white pines, together with air sampling data for SO_2, Linzon established three long-term study zones around Sudbury, Ontario, Canada. These zones were: inner zone near the sources (720 square miles), intermediate and outer, or clear, air zone. In each zone he studied the growth, yield and survival of selected white pines from 1953 to 1963. Trees were measured in height and in diameter DBH (breast high). Measurements were made in 1953 and again in 1963. After allowing for tree losses due to causes other than SO_2, economic yield losses over time in terms of volume growth loss were calculated for each SO_2 zone.

B. Scots Pine

Scots pine (*Pinus sylvestris* L.) is considered to be moderately sensitive to SO_2.[38] Farrar *et al.*[15] determined the present distribution of Scots pine in the industrial Pennines of England. Extensive mapping showed that Scots pine was uncommon in the area, even when conditions were appropriate for it. This area also coincides with regions where elevated atmospheric SO_2 concentrations occur. Farrar *et al.* concluded that SO_2 may be limiting the distribution of Scots pine.

C. Hybrid Poplar

Under natural conditions, trembling aspen (*Populus tremuloides* Michx.) is a very sensitive and dependable bioindicator of ambient SO_2 (Fig. 28). Dochinger *et al.*[12] and Dochinger and Jensen[13] have made a number of crosses to create hybrid poplar cultivars with known sensitivities to SO_2 at both high and low SO_2 concentrations. Clones were selected from a poplar hybrid (*Populus deltoides* Bartr. cv. 'Angulata' × *P. trichocarpa* torr. and Gray). SO_2-tolerant and SO_2-sensitive clones were established. Clone 211 was quite sensitive to SO_2, while clone 207 was more tolerant to SO_2, at both high and low SO_2 concentrations. Field use of these genetically defined, closely related hybrid poplar clones may allow for accurate biomonitoring of ambient SO_2 with a deciduous tree.

D. Vegetation Analyses Near Sources

1. Sulfate analyses of leaves

Lihnell[37] mapped SO_2 effects on vegetation around a shale-oil works in Kvarntorp, Sweden. He estimated the sulfate content of birch and apple leaves and spruce needles from trees at varying distances from the source. Sulfur in the leaves was precipitated out as barium sulfate and optical density or turbity measurements were made of resulting suspensions. Using this data, he was able to make sulfate diagrams for birch leaves which were compared with diagrams of growth reduction in several conifers. He found that both sets of diagrams were similar in form. He concluded that sulfate analysis was a reliable method for determining SO_2 in the Kvarntorp area.

2. Tree bark acidity

Samples of tree bark were obtained from eight linden (*Tilia cordata* Mill.), ash (*Fraxinus excelsior* L.), oak (*Quercus robur* L.), hornbeam (*Carpinus betulus* L.), alder (*Alnus incana* L.), hazel (*Coryllus avellana* L.) and Scots pine (*Pinus sylvestris* L.) in national parks in Southern Poland at various times of the year.[21] Tree bark was found to be most acidic when samples were taken in early spring and least acidic when samples were taken in summer. The pH of tree bark was highest in samples taken from the low SO_2 areas. Scots pine bark had the lowest pH readings and ash bark the highest.

The following simple protocol can be used to determine tree bark acidity:

(1) Use ash or linden trees for sampling. They are the least acidic and have rough surface textures.
(2) Bark samples should be taken in the early spring when they are the most acidic.
(3) External bark portions are the most acidic and they should be used for sampling.
(4) Comminute bark samples in distilled water and then read with a pH meter and a glass electrode.

Grodzinska[21] was able to correlate tree bark acidity and known SO_2 concentrations in air and rainfall to determine which forests in which parks were under significant stress caused by SO_2.

Johnsen and Sochting[29] examined changes in the bark properties of maple, ash, linden and elm trees near Copenhagen, Denmark. As the distance of the trees from the city center decreased, they found that the pH of bark samples decreased from 5.0 to 3.0 and the average sulfur content increased from 0.25 to 0.45‰.

3. Species richness and diversity

Rosenberg *et al.*[47] examined vegetation near a coal-burning power plant. In a mixed oak forest with Eastern white pine and hemlock, they found that species diversity and importance values for certain trees were inversely related with distance from the power plant. Differences with distance were more significant downwind from the plant than upwind from it. Diversity and importance of sweet birch (*Betula lenta* L.) and white pine increased with distance from the SO_2 source. The importance of white and red oak decreased with distance from the SO_2 source. Species richness and diversity increased as distance from the power plant increased. Rosenberg *et al.* felt that species richness and diversity were better indicators of SO_2 effects than were assessments of effects on individual trees alone.

IV. BIOMONITORING SO_2 WITH LOWER PLANTS

A. Mosses and Lichens

Mosses and lichens are primitive lower plants that are widely distributed. Both mosses and lichens lack vascular tissues and roots. All nutrients are taken up directly by the thallus from aqueous solutions. Lichens, in particular, are very efficient at taking up sulfur from the air and storing sulfur in excess of their needs.[17,44] Gilbert[17] demonstrated this when he determined the sulfur content of the lichen *Parmelia saxatilis* along a transect from Newcastle upon Tyne (Table 12).

Lichens have been widely used as SO_2 biomonitors. Lichens are symbiotic associations consisting of a fungus (the mycobiont), which is either an Ascomycete or Basidiomycete, and an alga (the phycobiont).[23,25] More advanced lichens have an upper and lower cortical layer, consisting of tightly woven hyphae of the fungus and an inner medulla layer, consisting of loosely woven hyphae and the algal symbiont. Lichens have a wide variety of shapes and forms. Growth forms of common lichens are summarized in Table 13. Foliose and fruticose lichens are the first to disappear when SO_2 concentrations rise. The algal symbiont appears to be the most sensitive part of the association with regard to SO_2 effects.[31,42]

There are a number of advantages and disadvantages to using epiphytic lichens and mosses as SO_2 biomonitors. These are summarized in Table 14. The advantages seem to outweigh the disadvantages and such lichens and mosses have been used extensively to determine SO_2 in the atmosphere. They are often used in conjunction with epiphytic mosses as both are found on the bases and trunks of trees.

The techniques used to evaluate epiphytic mosses and lichens in the field and laboratory for sensitivity to SO_2 are many and varied. Some of these

TABLE 12

ACCUMULATION OF SULFUR FROM THE ATMOSPHERE BY THALLI OF THE LICHEN *Parmelia saxatilis* ALONG A TRANSECT FROM NEWCASTLE UPON TYNE

Distance from center (km)	*Thallus sulfur content (ppm)*	*Sulfur in air (ppm)*
6·4	2870	0·02
13·6	695	0·014
33·6	225	—

Adapted from: Gilbert.[17]

TABLE 13

GROWTH FORMS OF COMMON LICHENS

Leprose	Powdery, little structure.
Crustose	Crust-like, may have little cortical tissue, intimate substrate contact.
Squamulose	Crust-like, margins becoming raised.
Placodioid	Crust-like, margins furrowed and lobe-like.
Foliose	Leaf-like thallus with distinct lower cortex and rhizinial attachment.
Fruticose	Erect hair-like or shrubby forms.

Adapted from references 23 and 25.

TABLE 14

SOME ADVANTAGES AND DISADVANTAGES OF USING EPIPHYTIC LICHENS AS BIOMONITORS OF ATMOSPHERIC SO_2

Advantages	*Disadvantages*
1. Lichens are very slow growing and long-lived.	1. Lichens have poor regenerative ability and may die when exposed to SO_2.
2. Lichens are easily handled and transplanted.	2. Lichens also accumulate fluoride and heavy metals as well as sulfur.
3. Lichens have no vascular system and readily absorb and accumulate sulfur from aqueous solutions.	3. Lichens do not respond quickly to high SO_2 concentrations.
4. Lichen species range in sensitivity to SO_2 from very sensitive to very resistant.	4. Counting lichen species in the field is laborious.
5. Lichens are more sensitive to low SO_2 concentrations than are higher plants.	5. Specialized knowledge of lichens is often needed.
6. There is good correlation between lichen species distribution and ambient SO_2 concentrations.	

Adapted from references 1, 18, 19, 22, 24, 29, 31 and 51.

TABLE 15

SOME CRITERIA USED TO EVALUATE EPIPHYTIC MOSSES AND LICHENS IN THE FIELD AND LABORATORY FOR SENSITIVITY TO SO_2

Criteria	*Methods*
1. Biomass	Dry weight
2. Chlorophyll content	Extraction, spectrophotometry
3. Color changes	Color photography
4. Degree of cover of each species	Visual estimation, tracings, planimetry
5. Fertility and reproduction	Check protonemata of mosses and soredia and isidia in lichens
6. Frequency of each species	Counts on trees
7. Growth and development	Photography, tracings, planimetry
8. Luxuriance	Visual estimation, biomass
9. Thallus injury	Visual estimation, infrared photography, planimetry of injured and healthy areas of thalli
10. Sulfur content	Chemical analysis
11. Total number of species	Counts on trees

Adapted from references 18, 23, 25, 34, 35, 50 and 51.

are summarized in Table 15. Whilst there seems to be no universal criterion of evaluation, the following are often used by investigators in field investigations:

(1) Determination of the total number of species.
(2) Determination of the degree of cover of each species.
(3) Determination of the frequency of each species.
(4) Determination of the luxuriance of each species.

1. Mapping studies

In addition to accumulating sulfur from sulfur dioxide, lichens and mosses are affected by SO_2 in another way. Tree bark can absorb sulfur from dilute solutions and, depending upon the nature of the bark and its innate buffering capacity, the pH of the bark may be decreased by accumulated sulfur. This has been shown for a number of tree species as distance from city centers decreases.[21,29] The pH change of tree bark leads to pronounced changes in the number of species present and a definite succession of species occurs as acid-sensitive ones are eliminated. A generalized scheme for the sequence of events in the loss of corticolous lichens from tree bases and trunks, as affected by SO_2, is given in Table 16. A number of

TABLE 16

SEQUENCE OF EVENTS IN THE LOSS OF CORTICOLOUS LICHENS FROM TREE BASES AND TRUNKS SUBJECTED TO MODERATE SO_2 POLLUTION (ANNUAL AVERAGE OF 30 TO 170 $\mu g/m^3$ SO_2)[a]

First loss of lichens
Lichens will be lost first from trees with acidic bark and low buffering capacity (e.g. birch and conifers)
Next loss of lichens
Sensitive lichens will be lost next from the bark of intermediate trees, such as oak and sycamore. As bark becomes acidified, lichen flora will gradually change to SO_2-resistant species.
Last to lose lichens
Trees with alkaline bark and high buffering capacity (e.g. elm) will be the last to lose their lichens.

Adapted from: Hawksworth and Rose.[25]

[a]The useful limits for corticolous lichens are between 30 and 170 $\mu g/m^3$ SO_2—as annual averages.[25]

investigators have followed similar changes on trees in the field and combined this with data on SO_2 concentrations obtained from mechanical monitoring devices to construct maps of areas where SO_2 is a problem.

To determine the relative sensitivities of lichens and mosses to SO_2, it is necessary to plot the distribution of individual species, growing on a standard substrate, in a given area or at a monitoring station, where ambient SO_2 is mechanically measured. Le Blanc and De Sloover[33] developed a numerical index value to express the extent of cover and frequency of occurrence for each lichen species they observed at SO_2 monitoring stations near Montreal, Canada. Their scheme is summarized in Table 17.

Transects are often followed from regions of high SO_2 concentrations to relatively non-polluted or clean areas. These transects may be very long, ranging up to 40 km in length and more. Epiphytic mosses and lichens are determined and evaluated at points along the transect. Different species predominate at different distances along the transect as it moves from the polluted area outward.

Gilbert[18] followed the incidence of an SO_2-sensitive lichen, *Evernia*

TABLE 17

NUMERICAL VALUES FOR EXTENT OF COVERAGE AND FREQUENCY OF OCCURRENCE FOR MOSSES AND LICHENS ON BASES AND TRUNKS OF ELM (*Ulmus americana*), ASH (*Fraxinus pennsylvanica*), POPLAR (*Populus canadensis*) OR MAPLE (*Acer saccharum*) AS DETERMINED BY LE BLANC AND DE SLOOVER[33]

Numerical value	*Extent of coverage and frequency*
1	Very rare, very low degree of coverage
2	Very infrequent, low degree of coverage
3	Infrequent, medium degree of coverage on some trees
4	Frequent, high degree of coverage
5	Very frequent, high degree of coverage on most trees

TABLE 18

CHANGES IN PER CENT COVER OF THE LICHENS *Lecanora conizaeoides* (SO_2 -RESISTANT) AND *Evernia prunastri* (SO_2 -SENSITIVE) ON ASH TREES (*Fraxinus excelsior*) ALONG A TRANSECT MOVING AWAY FROM NEWCASTLE UPON TYNE TO THE WEST

Lichens	*Per cent cover at various distances (in km) from Newcastle center*					
	8	*12*	*16*	*20*	*24*	*32*
L. conizaeoides	10	90	60	20	20	10
E. prunastri	0	5	10	25	25	40

Adapted from Gilbert.[18]

prunastri, and an SO_2-resistant lichen, *Lecanora conizaeoides*, on ash trees (*Fraxinus excelsior*) along a 32-km transect from Newcastle upon Tyne to the west. Incidence and per cent cover were used as criteria. *E. prunastri* began to appear around 12 km from the city center and increased gradually to 40% cover at the clean zone, 32 km from the city center. *L. conizaeoides*, however, increased in cover and vigor as the center of town was approached and decreased in cover with distance from the town center (Table 18). The thallus of *L. conizaeoides* is difficult to wet as it is coated with large amounts of fumarprotocetaric acid[25] and this makes it very resistant to SO_2 and an excellent indicator species.

Zones can be established which relate to the predominance of epiphyte species or growth forms along transects. These zones can be correlated with measured SO_2 concentrations and the relationship of epiphyte distribution and predominance to SO_2 concentrations (mean annual concentrations) can be determined.[19] A simplified zone scheme[25] for lichens on trees may have only three zones:

Zone 1	Inner zone	No foliose or fruticose lichens.
Zone 2	Intermediate or transition zone	
	(a) Inner zone	Foliose lichens present but poorly developed.
	(b) Outer zone	Fruticose lichens present but poorly developed.
Zone 3	Normal or outer zone	Fruticose and foliose lichens common and well developed.

Hawksworth and Rose[24] developed a more complex ten-zone scheme for use along SO_2 gradients in England and Wales. Once zones are established, sites along the transects are assigned to the appropriate zone and symbols can be plotted on a map, or contour lines can be constructed by linking identical sites (isopleths).

The establishment of zones of known SO_2 concentrations, or isopleths, makes possible the selection of indicator epiphyte species. Johnsen and Sochting[29] found a high degree of coincidence between the inner limits of lichen species distribution and SO_2 isopleths around Copenhagen, Denmark. They found three common lichen indicator species whose distribution corresponded to known annual average SO_2 concentrations (Table 19). Using both lichens and mosses, Gilbert[18] was able to determine average annual SO_2 concentrations around Newcastle upon Tyne that limit the occurrence of indicator species (Table 20).

An ambitious and extensive mapping project was used by British school children to determine comparative zones of SO_2 pollution on a national basis. Gilbert[20] reported the scheme of evaluation, which is summarized in Table 21. The children were asked to survey an area and score it for seven conspicuous and common lichens, one moss and one alga. Color photographs of the lichens were provided in a booklet given to the children. The

TABLE 19

INDICATOR LICHEN SPECIES IN RELATION TO AVERAGE ANNUAL ATMOSPHERIC SO_2 CONCENTRATIONS IN THE COPENHAGEN AREA, AS DETERMINED BY JOHNSEN AND SOCHTING[29]

Indicator lichens	*Corresponding annual average SO_2 concentrations ($\mu g/m^3$)*
Buellia punctata	90–110
Lecanora subfusca	70– 80
Physcia pulverulenta	40

TABLE 20

AVERAGE ANNUAL ATMOSPHERIC CONCENTRATIONS THAT LIMIT THE OCCURRENCE OF INDICATOR LICHENS AND MOSSES AROUND NEWCASTLE UPON TYNE, AS DETERMINED BY GILBERT[18]

Species	*Limiting SO_2 concentrations (in ppm)*
Lichens	
Paramelia saxatalis / *Paramelia fuliginosa*	0·020
Mosses	
Grimmia pulvinata / *Hypnum cupressiforme*	0·016

indicator species were observed on deciduous trees, as well as acid and alkaline stone, and recorded on the score sheet. Seven lichen zones were used in the evaluation, each corresponding to a different SO_2 level. When the score sheets were collected, a lichen zone map (at 1:63360) of the area surveyed was prepared. By fitting all the areas surveyed together, a lichen zone map of all of Great Britain was obtained. As these zones relate to known SO_2 concentrations, regions of the country with high SO_2 pollution problems could be determined.

2. Transplant studies

Brodo[4] developed a method of transplanting lichens from trees growing in clean areas to areas where air pollution was a problem. Circular discs of bark, with lichens on them, were cut from healthy trees and moved without

TABLE 21

THE SEVEN ZONES OF THE SCHOOLCHILDREN'S AIR POLLUTION SURVEY OF GREAT BRITAIN

Zones	Indicator cryptogams present
0 Heavy pollution (near source or city center)	No lichens, only the alga *Pleurococcus* on trees and acid stone
1 Less pollution	Grey–green crusty lichen *Lecanora conizaeoides* on tree bases and acid stone
2 Less pollution	Orange leafy lichen *Xanthoria parietina* on alkaline stone
3 Less pollution	Grey leafy lichen *Parmelia* on acid stone, moss *Grimmia pulvinata* on alkaline stone, *Lecanora* and *Pleurococcus* on trees
4 Less pollution	Grey leafy lichens begin on tree bases
5 Less pollution	*Lecanora* becomes less common, shrubby lichens appear, especially *Evernia*
6 Very clean air	Shrubby lichens common, especially *Usnea*

Adapted from Gilbert.[20]

difficulty. Subsequent development of transplanted lichens was determined by comparing periodic photographs.

Le Blanc and Rao[32,34] used the Brodo technique in transplanting mosses and lichens around Sudbury, Ontario, Canada. These were continuously evaluated for one year. The five zones of SO_2 concentrations around Sudbury, established by Dreisinger and McGovern[14], were used to evaluate the transplanted epiphytes:

Zones	*Corresponding annual average SO_2 concentration in ppm*
I	0·30–0·42
II	0·20–0·30
III	0·01–0·20
IV	0·005–0·01
V	Less than 0·005

On a long-term basis, they were able to conclude that when SO_2 concentrations were below 0·002 ppm, no injury to transplanted mosses or lichens would occur (Table 22). In the zones nearest Sudbury most of the transplanted mosses and lichens were dead or extensively damaged. Changes noted in the most common epiphytes used are summarized in Table 23.

Lichen transplants have been used extensively in Europe, especially by Schönbeck[49,50] in Germany. *Hypogymnia physodes* on bark samples were placed on boards at sampling locations all over Germany. Ten lichen

TABLE 22
EFFECTS OF LONG RANGE CONCENTRATIONS OF SO_2 ON EPIPHYTIC MOSSES AND LICHENS AROUND SUDBURY, ONTARIO, CANADA AS DETERMINED BY LE BLANC AND RAO[34]

Average annual SO_2 concentrations	*Effects on epiphytic mosses and lichens*
Above 0·03 ppm	Acute injury
Between 0·006 and 0·03 ppm	Chronic injury
Below 0·002 ppm	No injury

TABLE 23
SOME CHANGES IN TRANSPLANTED MOSSES, *Orthotrichum obtusifolium* AND *Pylaisiella polyantha*, AND LICHENS, *Parmelia sulcata* AND *Physcia millegrana*, CLOSE TO SO_2 SOURCES IN SUBDURY, ONTARIO, CANADA, AS DETERMINED BY LE BLANC AND RAO[34]

Features evaluated	*Changes in mosses and lichens close to SO_2 sources*
Color	Lichens changed from grey–green to white–brown Mosses changed from green to golden brown
External features	Lichen cortices covered with waxy substance, no soredia on *P. sulcata*
Internal features	Moss leaf cells bleached Death and plasmolysis of algal cells in lichens Sulfur content increased for both epiphytes

samples were used on each board, to allow for statistical analyses of data obtained at each site. Where the average SO_2 concentration was 0·23 mg/m^3, *H. physodes* died in 29 days. In low SO_2 areas, where SO_2 concentrations averaged 0·08 mg/m^3, 60% necrosis of the thalli of *H. physodes* occurred after 68 days. Damage was determined by means of infrared photography and planimetric measurements of healthy and damaged thallus surface areas. Schönbeck concluded that *H. physodes* is a good indicator for low concentrations of atmospheric SO_2.

B. Microorganisms

1. Field observations of plant diseases

In the course of a number of investigations on SO_2 effects on vegetation near sources, it has been observed that the presence or absence of certain plant diseases, mostly on leaves, can serve as an indication of SO_2 pollution. This is not unexpected, as sulfur is known to be a very effective fungicide.

Changes in the frequency and distribution of fungal leaf diseases in relation to SO_2 concentrations are usually noted in the inner zone and then the intermediate zone with regard to distance from SO_2 sources. Scheffer and Hedgcock[48] made extensive observations on needle and leaf diseases on deciduous and coniferous trees around the Washoe smelter at Anaconda, Montana, USA. They found that many rust and leafspot fungi

TABLE 24

PLANT DISEASE FUNGI ADVERSELY AFFECTED BY ATMOSPHERIC SO_2 UNDER FIELD CONDITIONS

Fungi	*Hosts*
Cronartium commandrae Pk.	Lodgepole pine
Cronartium harknessii Mein.	Lodgepole pine
Cronartium ribicola Fisch.	Eastern white pine
Hypodermella laricis Tub.	Larch
Lophodermium juniperum (Fr.) de N.	Common juniper
Lophodermium pinastri (Schrad.) Chev.	Ponderosa pine
Melampsora albertensis Arth.	Quaking aspen
Melampsora occidentalis Jacks.	Black cottonwood
Microsphaera alni	Lilac, oak
Pucciniastrum pustulatum (Pers.) Diet.	Grand and subalpine firs
Rhytisma acerinum (Pers.) Fries.	Maple, sycamore maple

Adapted from references 1, 3, 26, 30, 39, 48 and 53.

in the genera *Cronartium, Coleosporium, Melampsora, Peridermium, Pucciniastrum, Lophodermium, Hypoderma*, and *Hypodermella* were either absent or inhibited in the SO_2 zone where tree injury was greatest (Table 24). Development of many diseases was also inhibited where SO_2 concentrations caused only moderate plant damage. Linzon[39] noted that the white pine blister rust fungus, *Cronartium ribicola*, was almost completely eliminated in the inner SO_2 zone around Sudbury, Ontario, Canada. The powdery mildew fungus, *Microsphaera alni*, has been shown to be a good indicator of high urban SO_2 concentrations on lilac[26] and on oak leaves near pulp mills.[30] *Lophodermium juniperum* is common on *Juniperus communis* in England, except in the high SO_2 areas around London.[1]

Trees weakened by SO_2 may be more susceptible to plant pathogenic fungi and insects, such as bark beetles. Scheffer and Hedgcock[48] found that root and collar rot of trees, caused by the opportunistic soil-borne fungus *Armillaria mellea* (Fr.) Quel., was more prevalent on SO_2-injured trees close to the source in Montana, USA.

2. *Mapping tar spot of maple*

Tar spot is a common disease of several maple species. The distinctive, raised, shiny, hard coal-black spots on upper leaf surfaces resemble spots

TABLE 25

RELATIONSHIP OF SEVERITY OF TAR SPOT DISEASE (CAUSED BY THE FUNGUS *Rhytisma acerinum* (Pers.) Fries) ON SYCAMORE MAPLE (*Acer pseudoplatanus* L.) TO ANNUAL AVERAGE CONCENTRATIONS OF ATMOSPHERIC SO_2, AS DETERMINED BY BEVAN AND GREENHALGH[3]

Severity of tar spot on maple	*Corresponding annual average SO_2 concentration in $\mu g/m^3$*
No tar spot, leaves clean	Greater than 85
Tar spots rare, one spot per tree or less[a]	70–85
Tar spots infrequent, average of ten spots per tree	55–70
Tar spots frequent, two or more spots per leaf common	40–55
Tar spots on large numbers of leaves—each with at least one spot, many with between one and four spots	25–40
Tar spots on all leaves, many spots per leaf	Less than 25

[a]Number of tar spots per 100 cm^2 leaf area.

FIG. 30. Comparison of the incidence of tar spot disease (caused by the fungus *Rhytisma acerinum* (Pers.) Fries) on maple (*Acer* spp.) as it would appear in relatively clean air (right), with many spots per leaf and as it would appear where SO_2 concentrations are high (left), with only one spot per leaf. (Courtesy: USDA.)

made by tar. Each spot is a stroma of the fungus *Rhytisma acerinum* (Pers.) Fries. SO_2 and sulfur in solution are fungicidal to *R. acerinum*.

Beven and Greenhalgh[3] and Vick and Bevan[53] surveyed tar spot incidence on sycamore maple, *Acer pseudoplatanus* L., in the greater Liverpool (Great Britain) area. They found that they could correlate severity of tar spot on sycamore maple with corresponding annual average concentrations of SO_2 (Table 25). Examples of differences in severity of tar spot incidence are illustrated in Fig. 30. These can be compared with the criteria for tar spot severity, as given in Table 25.

3. Incidence of airborne bacteria

Micro-organisms, particularly fungal spores and bacterial cells, are common in ambient air, especially when relative humidity exceeds 75 per cent. The possibility of using changes in counts of airborne micro-organisms to monitor SO_2 concentrations in ambient air has been much discussed, but not extensively explored.

In laboratory experiments, Lighthart *et al.*[36] found that SO_2 could affect the death rate of cells in aerosols of the pigmented cosmopolitan bacterium *Serratia marcescens*. An abrupt decrease in the number of viable cells in an aged aerosol of *S. marcescens* was observed after exposure to 3·75 mg of SO_2 per m^3 at 80 per cent relative humidity.

REFERENCES

1. BARKMAN, J. J., F. ROSE and V. WESTHOFF. (1969). The effects of air pollution on non-vascular plants. *Proc. European Congress on the Influence of Air Pollution on Plants and Animals, Wageningen*, pp. 237–41.
2. BARRETT, T. W. and H. M. BENEDICT. (1970). Sulfur dioxide. In: *Recognition of Air Pollution Injury to Vegetation: A Pictorial Atlas.* (Jacobson, J. S. and Hill, A. C. (eds.)). Air Poll. Contr. Assoc., Pittsburgh, Pa., pp. C1–C17.
3. BEVAN, R. J., and G. N. GREENHALGH. (1976). *Rhytisma acerinum* as a biological indicator of pollution. *Environ. Pollut.* **10**: 271–85.
4. BRODO, I. M. (1961). Transplant experiments with corticolous lichens using a new technique. *Ecology*, **42**: 838–41.
5. CONARD, H. S. (1956). *The Mosses and Liverworts.* Wm. C. Brown Co., Dubuque, Iowa. 226 pp.
6. DARLEY, E. F. (1960). Use of plants for air pollution monitoring. *J. Air Poll. Contr. Assoc.*, **10**: 198–9.
7. DAVIS, D. D. and R. G. WILHOUR. (1976). *Susceptibility of woody plants to sulfur dioxide and photochemical oxidants.* US EPA Publication 600/3–76–102. 71 pp.
8. DIXON, H. N. (1954). *The Student's Handbook of British Mosses.* Sumfield & Day, Ltd, London. 582 pp.
9. DOCHINGER, L. S. and C. E. SELISKAR. (1965). Results from grafting chlorotic dwarf and healthy Eastern white pine. *Phytopathology*, **55**: 404–7.
10. DOCHINGER, L. S. and C. E. SELISKAR. (1970). Air pollution and the chlorotic dwarf disease of Eastern white pine. *For. Sci.*, **16**: 46–55.
11. DOCHINGER, L. S., F. W. BENDER, F. L. FOX and W. W. HECK. (1970). Chlorotic dwarf of Eastern white pine caused by an ozone and sulphur dioxide interaction. *Nature (London)*, **225**: 476.
12. DOCHINGER, L. S., A. M. TOWNSEND, D. W. SEEFRIST and F. W. BENDER. (1972). Responses of hybrid poplar trees to sulfur dioxide fumigation. *J. Air Poll. Contr. Assoc.* **22**: 369–71.
13. DOCHINGER, L. S. and K. F. JENSEN. (1975). Effects of chronic and acute exposure to sulfur dioxide on the growth of hybrid poplar cuttings. *Environ. Pollut.*, **9**: 219–29.
14. DREISINGER, M. F. and P. C. MCGOVERN. (1970). Monitoring atmospheric sulphur dioxide and correlating its effects on crops and forests in the Sudbury area. In: *Impact of Air Pollution on Vegetation Conference.* (Linzon, S. N. (ed.)), Air Poll. Contr. Assoc., Pittsburgh, Pa. pp. 11–28.
15. FARRAR, J. F., J. RELTON and A. J. RUTTER. (1977). Sulphur dioxide and the scarcity of *Pinus sylvestris* in the industrial Pennines. *Environ. Pollut.* **14**: 63–68.
16. FERRY, B. W., M. S. BADDELEY and D. L. HAWKSWORTH. (1973). *Air Pollution and Lichens.* Athlone Press, Univ. London. 389 pp.
17. GILBERT, O. L. (1965). Lichens as indicators of air pollution in the Tyne valley. In: *Ecology and the Industrial Society*, (Goodman, G. T., Edwards, R. W. and Lambert J. M. (eds.)), Blackwell Scientific, Oxford, pp. 35–47.
18. GILBERT, O. L. (1969). The effect of SO_2 on lichens and bryophytes around

Newcastle upon Tyne. *Proc. European Congress on the Influence of Air Pollution on Plants and Animals, Wageningen*, pp. 223–36.

19. Gilbert, O. L. (1970). A biological scale for the estimation of sulphur dioxide pollution. *New Phytol.*, **69**: 629–34.
20. Gilbert, O. L. (1974). An air pollution survey by school children. *Environ. Pollut.*, **6**: 175–80.
21. Grodzinska, K. (1977). Acidity of tree bark as a bioindicator of forest pollution in Southern Poland. *Water, Air, and Soil Pollution*, **8**: 3–7.
22. Guderian, R. (1977). *Air Pollution: Phytotoxicity of Acidic Gases and its Significance in Air Pollution Control*, Springer-Verlag, New York 127 pp.
23. Hale, Jr., M. E. (1974). *The Biology of Lichens*. Edward Arnold Pub., London, 181 pp.
24. Hawksworth, D. L. and F. Rose. (1970). Qualitative scale for estimating sulphur dioxide air pollution in England and Wales using epiphytic lichens., *Nature (London)*, **227**: 145–8.
25. Hawksworth, D. L. and F. Rose. (1976). *Lichens as Pollution Monitors*. Edward Arnold Pub., London, 59 pp.
26. Hibben, C. R. and M. P. Taylor, (1975). Ozone and sulfur dioxide effects on the lilac powdery mildew fungus. *Environ. Pollut.* **9**: 107–14.
27. Hill, A. C. (1971). Vegetation: A sink for atmospheric pollutants. *J. Air Poll. Contr. Assoc.*, **21**: 341–6.
28. Houston, D. B. (1974). Response of selected *Pinus strobus* L. clones to fumigations with sulfur dioxide and ozone. *Can. J. For. Res.*, **4**: 65–68.
29. Johnsen, I. and U. Sochting. (1973). Influence of air pollution on the epiphytic lichen vegetation and bark properties of deciduous trees in the Copenhagen area. *Oikos*, **24**: 344–51.
30. Köck, G. (1935). Eichenmehltau und Rauch gass-schaden. *A. Pfl. Krankh.* **45**: 44–45.
31. Le Blanc, F. (1969). Epiphytes and air pollution. *Proc. European Congress on the Influence of Air Pollution on Plants and Animals, Wageningen*, pp. 211–22.
32. Le Blanc, F. and D. N. Rao. (1966). Reaction de quelques lichens et mousses epiphytiques a l' anhydride sulfureux dans la region de Sudbury, Ontario. *Bryologist*, **69**: 338–46.
33. Le Blanc, F., and J. De Sloover. (1970). Relation between industrialization and the distribution and growth of epiphytic lichens and mosses in Montreal. *Can. J. Bot.*, **48**: 1485–96.
34. Le Blanc, F., and D. N. Rao. (1973). Effects of sulphur dioxide on lichen and moss transplants. *Ecology*, **54**: 612–17.
35. Le Blanc, F., and D. N. Rao. (1975). Effects of air pollutants on lichens and bryophytes. *Responses of Plants to Air Pollution*, (Mudd, J. B. and Kozlowski T. T. (eds.)), Academic Press, N. Y. pp. 237–272.
36. Lighthart, B., V. E. Hiatt and Q. T. Rossano, Jr. (1971). The survival of airborne *Serratia marcescens* in urban concentrations of sulfur dioxide. *J. Air Poll. Contr. Assoc.*, **21**: 639–42.
37. Lihnell, D. (1969). Sulphate content of tree leaves as indicator of SO_2 air pollution in industrial areas. *Proc. European Congress on the Influence of Air Pollution on Plants and Animals, Wageningen*, pp. 341–52.
38. Linzon, S. N. (1969). Symptomatology of sulphur-dioxide injury on vege-

tation. In: *Handbook of Effects Assessment: Vegetation Damage*. (Lacasse, N. L. and Moroz, W. J. (eds.)), Center for Air Environment Studies, Penn. State Univ., Univ. Park, Pa, Chapt. VIII.

39. LINZON, S. N. (1971). Economic effects of sulphur dioxide on forest growth. *J. Air Poll. Contr. Assoc.*, **21**: 81–86.
40. LINZON, S. N., W. D. MCILVEEN and P. J. TEMPLE. (1973). Sulphur dioxide injury to vegetation in the vicinity of a sulphite pulp and paper mill. *Water, Air and Soil Poll.*, **2**: 129–34.
41. LINZON, S. N., P. J. TEMPLE and R. G. PEARSON. (1979). Sulfur concentrations in plant foliage and related effects. *J. Air Poll. Contr. Assoc.*, **29**: 520–25.
42. NIEBOER, E., D. H. S. RICHARDSON, K. J. PUCKETT and F. D. TOMASSINI. (1976). The phytotoxicity of sulphur dioxide in relation to measureable responses in lichens. *Effects of Air Pollutants on Plants*. (Mansfield, T. A. (Ed.)), Cambridge Univ. Press, pp. 61–85.
43. PILEGAARD, K. (1978). Airborne metals and SO_2 monitored by epiphytic lichens in an industrial area. *Environ. Pollut.*, **17**: 81–92,
44. PYATT, F. B. (1973). Plant sulphur content as an air pollution gauge in the vicinity of a steelworks. *Environ. Pollut.*, **5**: 103–15.
45. ROBERTS, B. R. (1974). Foliar sorption of atmospheric sulfur dioxide by woody plants. *Environ. Pollut.*, **7**: 133–40.
46. ROBERTS, B. R. (1976). The response of field-grown white pine seedlings to different sulphur dioxide environments. *Environ. Pollut.*, **11**: 175–80.
47. ROSENBERG, C. R., R. J. HUTNIK and D. D. DAVIS. (1979). Forest composition at varying distances from a coal-burning power plant. *Environ. Pollut.* **19**: 307–17.
48. SCHEFFER, T. C. and G. C. HEDGCOCK. (1955). *Injury to northwestern forest trees by sulfur dioxide from smelters*. USDA Forest Service Tech. Bul. No. 1117, 49 pp.
49. SCHÖNBECK, H. (1968). Einfluss von Luftveruneinigungen (SO_2) auf transplantierte Flechten. *Naturwissenschaften*, **55**:451–2.
50. SCHÖNBECK, H. (1969). Eine Method zur Erfassung der biologischen Wirkung von Luftveruneinigungen durch transplantierte Flechten. *Staub Reinhalt.* Luft. **29**: 1.
51. SKYE, E. (1979). Lichens as biological indicators of air pollution. *Ann. Rev. Phytopathol.*, **17**: 325–41.
52. VAN RAAY, A. (1969). The use of indicator plants to estimate air pollution by SO_2 and HF. *Proc. European Congress on the Influence of Air Pollution on Plants and Animals, Wageningen*, pp. 319–28.
53. VICK, C. M. and R. BEVAN. (1976). Lichens and tar spot fungus (*Rhytisma acerinum*) as indicators of sulfur dioxide pollution on Merseyside. *Environ. Pollut.*, **11**: 203–16.

CHAPTER 6

Biomonitoring Hydrogen Fluoride

I. INTRODUCTION

Atmospheric fluoride may occur as the gas, hydrogen fluoride (HF), as particulate fluoride, or as HF adsorbed onto other particulates. HF is the most phytotoxic form of fluoride and it has been studied extensively. As a result, it is the only form of fluoride that will be considered here. For convenience, HF will be referred to simply as fluoride.

Like sulfur from SO_2, fluoride accumulates in higher plants, at the tips and margins of leaves. When threshold concentrations are exceeded, tissue necrosis occurs, from the tips of leaves back to the bases, along margins, or both. These symptoms of acute fluoride injury occur on sensitive plants near point sources and diminish with distance from the source. Effects on vegetation can be mapped. The shape of the affected area reflects the extent and duration of the fluoride source, prevailing winds and topography.

Fluoride concentrations in the atmosphere, soil and vegetation can be determined by a variety of chemical and mechanical methods. Data from chemical and mechanical analyses can be combined with responses of sensitive and resistant plants in an area to determine an index or estimation of the relative concentration of fluoride in that area. A number of higher plants and lichens are known to be sensitive or resistant to atmospheric fluoride[8,9,13,14,18,19] and they can be used as effective biomonitors (Table 26).

II. EVALUATING FLUORIDE INJURY

A. Symptoms

A number of broadleaved and coniferous plants respond to known atmospheric fluoride concentrations in predictable ways.[1,18] Symptoms of

TABLE 26

SOME PLANTS REPORTED TO BE SENSITIVE OR RESISTANT TO HYDROGEN FLUORIDE UNDER FIELD CONDITIONS

Very sensitive
Chinese apricot (*Prunus armeniaca* L.)
Douglas fir (*Pseudotsuga menziesii* (Mirb.) Franco)
Freesia (*Freesia hybride* Hort.)
Gladiolus (*Gladiolus hortulanus* Bailey)
Pines
Scots (*Pinus sylvestris* L.)
Western yellow (*Pinus ponderosa* Laws.)
Spruces
Engelmann (*Picea engelmanni* Parry ex. Engelm.)
Norway (*Picea abies* L.)
Serbian (*Picea omorika* L.)
Sitka (*Picea sitchensis* (Bong.) Carr)
Tolerant to resistant
European larch (*Larix decidua* Mill.)
Firs
Noble (*Abies procera* Rehd.)
White (*Abies concolor* (Gord. & Glend.) Hoopes)
Western hemlock (*Tsuga heterophylla* (Raf.) Sarg.)

Adapted from references 2, 14, 18 and 19.

fluoride injury on broadleaved plants usually include marginal and/or terminal necrotic areas, with sharply defined, pigmented margins (Figs 31 and 32). This necrotic tissue may separate and fall, but defoliation usually does not occur. On conifers, fluoride-caused necrosis begins at the tips of the current season's needles and progresses back to needle bases. The color of injured areas changes from green to yellow to red – brown. Extensively burned needles fall off (Fig. 33). Newly emerging and young needles are sensitive to fluoride. Older needles are not injured.

There are a number of advantages to using plant responses to biomonitor fluoride (Table 27). If known fluoride-sensitive and resistant plants are present in an area, the incidence and severity of characteristic fluoride injury symptoms on the sensitive plants can provide an index or measure of the presence of fluoride at phytotoxic concentrations. This acute injury can be quantified by determining per cent foliar injury in the field by visual estimation.[8, 14, 19] Laboratory methods include tracing outlines of injured leaf areas with a planimeter[16] or using a planimeter to trace photocopies of injured leaf areas.[20] Symptoms can also be reduced to

Fig. 31. Acute injury on tips and margins of gladiolus leaves (*Gladiolus hortulanus* Bailey), caused by hydrogen fluoride (HF). (Courtesy: USDA.)

Fig. 32. Acute injury on poplar leaves (*Populus* sp.), caused by hydrogen fluoride (HF). (Courtesy: USDA.)

FIG. 33. Acute injury on new growth of blue spruce (*Picea pungens* Engelm.), caused by hydrogen fluoride (HF). All of the needles on the current season's growth have fallen. (Courtesy: USDA.)

numerical indices to quantitatively express the response of individual biomonitor plants or groups of plants at sampling or monitoring locations.[2]

As with other air pollutants, the use of symptoms alone to determine fluoride in the atmosphere is not without problems (Table 27). Symptom expression in response to fluoride dose (concentration × time) does not occur until enough plant growth and fluoride accumulation are accom-

TABLE 27

COMPARATIVE ADVANTAGES AND DISADVANTAGES OF USING PLANTS TO BIOMONITOR FLUORIDE

Advantages

(1) There are many higher plants and lichens that are sensitive to fluoride.
(2) Characteristic symptoms occur on sensitive plants at known fluoride concentrations.
(3) Fluoride concentrations can be accurately determined in plants.
(4) Advantages (2) and (3) can be correlated.
(5) Symptoms usually increase as fluoride concentrations in vegetation and air increase.

Disadvantages

(1) Sensitive plants must be grown in a fluoride-free greenhouse and then be transported to the field.
(2) Response of sensitive plants to fluoride may be slow. Plants must grow, accumulate fluoride and then show symptoms.
(3) Symptoms may not reflect fluoride dose (concentration $\times$ time):
    (a) Rain may wash some fluoride away.
    (b) Some fluoride may be converted to less harmful forms by plants.
    (c) Plant nutritional status may affect symptom expression.
(4) Fluoride symptoms may be mimicked by:
    (a) Moisture stress
    (b) Low temperatures
    (c) *Botrytis*

Adapted from: references 10–12, 15, 17–20.

plished. Even then, cool temperatures[18] and nutritional deficiencies[12] may delay or reduce symptom expression. Moisture stress, cold temperatures and the facultative parasite *Botrytis* may cause injury that resembles that caused by fluoride. As a result, symptom expression should be combined with fluoride analyses and, if possible, data on atmospheric fluoride concentrations, to conclusively demonstrate that plant symptoms relate to atmospheric fluoride concentrations in a given area.

B. Chemical Analyses

Fluoride is an element that normally occurs in plant tissues. Under normal conditions, many plant leaves accumulate 0·05 to 0·10 ppm fluoride during a growing season.[18] Background levels of fluoride must be considered when results from chemical analyses of plant parts are evaluated.

A number of methods for analyzing fluoride in plant tissues can be used.[1,4] Cooke *et al.*[4] have provided the most recent summary of methods.

TABLE 28
PREPARATION OF VEGETATION SAMPLES FOR FLUORIDE ANALYSES

Washing

Samples should be thoroughly washed to remove dust and other particulates.
(1) Wash several times in detergent: EDTA mixes.
(2) Rinse several times with deionized water.

Drying

Samples must be carefully dried: 60–80°C for 12–48 h, depending on the nature of the tissues.
Losses of volatile fluoride may occur if the temperature is too high or the drying time is too long.

Grinding

Samples must be reduced to a fine homogeneous powder. Screening through a 60 mesh sieve is suggested.

Adapted from: Cooke *et al.*[4]

The use of a fluoride ion-selective electrode is advocated for rapid, accurate routine fluoride determinations in vegetation.

The accurate quantitative determination of fluoride in plant samples depends upon careful preparation and handling of samples prior to chemical analysis (Table 28). Careful attention must be given to washing the samples. Dusts and fluoride-containing particulates must be removed. Drying at fairly low temperatures for defined time periods is necessary in order to prevent volatilization of fluorides in samples. Samples must be thoroughly mixed and ground to provide a uniform, homogeneous sample for analysis.

When data on symptom expression in plants is combined or correlated with fluoride analyses of those plants, it is then possible to use those plants as biomonitors of atmospheric fluoride and to determine the source, the rate and extent of a fluoride pollution problem.[19]

III. BIOMONITORING FLUORIDE WITH HIGHER PLANTS

A. Chinese Apricot

Chinese apricot (*Prunus armeniaca* L.) is very sensitive to low concentrations of fluoride.[5,18] It responds to fluoride in a manner similar to that of most other sensitive broadleaved deciduous trees.

Fluoride injury symptoms first develop along the margins of apricot leaves. Semicircular lesions of bleached necrotic tissue (1–3 cm in diameter) may coalesce until entire leaf margins are included. Symptoms of this type are shown in Fig. 32. A narrow, prominent, red–brown band separates necrotic from green leaf tissue. Necrotic tissue may fall out, but defoliation does not usually occur. Young, soft actively growing leaves are the most sensitive to fluoride. These are often found on sucker shoots, where active rank growth occurs.

B. Conifers

Many conifers are known to be sensitive to fluoride (Fig. 33). A number of reports have been made on the effects of atmospheric fluoride from aluminum smelters and fertilizer plants on various conifers.[1,2,14,17,18] Douglas fir and Ponderosa pine are two conifers that have been extensively evaluated as biomonitors of fluoride.

1. Douglas fir

Treshow *et al.*[17] established four transects moving away from a phosphate reduction plant. They determined the responses of ten Douglas fir trees (*Pseudotsuga menziesii* (Mirb.) Franco) at three study sites along each transect. The location of each study site on each transect was related to average fluoride content of tree needles. Group A (control) was the farthest from the fluoride source. Average fluoride concentrations of needles at Group A were 50 ppm or less. Group B was intermediate in distance from the source. Fluoride concentrations of needles at Group B ranged from 51 to 200 ppm. Group C (high) was near the source and average fluoride concentrations there were over 200 ppm. At each study site, radial growth and needle length and dry weight were determined before and during four years of operation of the phosphate plant. Increment borings were used to determine radial growth. Three shoots were collected from each tree. Average needle lengths and dry weight of 100 g needles were determined. Other needles were used for fluoride analyses.

Radial growth of Douglas fir was inhibited as average fluoride concentrations in needles increased, even when symptoms were not evident (Table 29). Suppression of radial growth, even without needle symptoms, was considered to be the most significant response of Douglas fir to fluoride. Treshow *et al.*[17] were able to determine the threshold for tree growth reduction without fluoride symptoms. This occurred when the average fluoride concentrations in needles ranged from 100 to 200 ppm (Table 30). Foliar symptoms appeared when fluoride exceeded 200 ppm and tree death occurred when needle fluoride concentrations exceeded 500 ppm.

TABLE 29

GROWTH RESPONSES OF DOUGLAS FIR (*Pseudotsuga menziesii* (Mirb.) Franco) TO ATMOSPHERIC FLUORIDE FROM A PHOSPHATE REDUCTION PLANT OVER A FOUR-YEAR PERIOD, AS DETERMINED BY TRESHOW *et al.*[17]

Growth responses	*Sampling locations*		
	Group A Under 50 ppm HF (Control)	*Group B 51–200 ppm HF (Intermediate)*	*Group C Over 200 ppm HF (High)*
Radial growth (mm)	1·47	1·16	1·10*
Needle length (mm)	26·3	24·3*	24·2*
Dry weight (100 g needles)	0·68	0·68	0·63

*Significantly different than control $P = 0{\cdot}01$.

TABLE 30

RESPONSES OF DOUGLAS FIRS (*Pseudotsuga menziesii* (Mirb.) Franco) TO ATMOSPHERIC FLUORIDE AS RELATED TO FLUORIDE CONTENT OF DOUGLAS FIR NEEDLES, AS DETERMINED BY TRESHOW *et al.*[17]

Responses of Douglas firs	*Corresponding fluoride concentrations in needles (in ppm)*
(1) Threshold for growth reduction without foliar symptoms.	100–200
(2) Foliar symptoms.	Over 200
(3) Trees killed.	Over 500

A distinct stimulation of Douglas fir needle length was a very early symptom of fluoride accumulation. Increasing fluoride concentrations, however, resulted in a reduction in average needle length (Table 29).

Reductions in dry weight of needles occurred when average needle concentrations of fluoride exceeded 200 ppm, and foliar burning occurred (Tables 29 and 30).

Douglas fir trees appear to be very sensitive biomonitors of fluoride. As perennial plants, long-term assessment of Douglas fir trees around fluoride sources would provide an excellent continuing index or measure of air quality with regard to atmospheric fluoride.

2. *Ponderosa pine*

Ponderosa pine (also known as western yellow pine) (*Pinus ponderosa* Laws.) is a native conifer in much of western North America. Adams *et al.*[2] have extensively investigated the effects of fluoride on Ponderosa pine seedlings near Spokane, Washington, USA.

Three study areas were established around the fluoride source. Atmospheric fluoride was monitored mechanically at each site. Five-gallon containers of soil, planted with four five-year-old Ponderosa pine seedlings, were placed at each study area. By periodically evaluating these seedlings, Adams *et al.* were able to quantify needle injury and calculate a needle burn injury index, an individual pine seedling injury index and location injury averages (Table 31). At each study area, a linear correlation was found between the average degree of needle injury and the average atmospheric fluoride concentration. This is one of a few cases in the literature where it has been possible to calculate such a linear correlation.

In other work with Ponderosa pine,[1] it was found that weekly applications of lime (calcium carbonate), with an appropriate chemical spreader and sticker, protected needles from fluoride injury. The lime reacted with airborne fluoride to produce calcium fluoride, which prevented fluoride entry into needles. By comparing lime-treated and non-treated trees, the effects of fluoride on tree response and development could be demonstrated.

TABLE 31

NUMERICAL CALCULATION OF FLUORIDE INJURY ON PONDEROSA PINE SEEDLINGS (*Pinus ponderosa* Laws.) AT FIELD SAMPLING LOCATIONS, AS DETERMINED BY ADAMS *et al.*[2]

Calculate needle burn index

(1) Use 10 per cent of the burned current season's needles as a sample.
(2) Measure needle lengths: burned and total

$$\text{Needle burn index} = \frac{\text{Avg. length of burned tissue} \times 100}{\text{Avg. needle length}}$$

Calculate individual pine injury index

$$\text{Pine injury index} = \frac{\text{Needle burn index} \times \%\text{ of needles burned}}{100}$$

Calculate location averages

Location average = average of all individuals at a location

C. Gladiolus and Freesia

Many varieties of gladiolus, freesia, tulip, iris, lily, narcissus and other monocots are known to be sensitive to an atmospheric fluoride.[2,3,10,11,15,16,19,20] Gladiolus (*Gladiolus hortulanus* Bailey) is the plant most widely used to biomonitor for fluoride.[11] Responses of this plant to fluoride have been extensively documented and a number of evaluation schemes have been developed.

1. Gladiolus

All parts of the gladiolus plant are affected by fluoride (Table 32). Leaf injury is the response that is used to assess fluoride concentrations. The extent and appearance of this leaf injury varies with the gladiolus variety and fluoride dose (concentration × time). In general, necrotic tip-burn occurs, beginning at the tips and extending downwards across the leaf or along the margins, or both (Fig. 31). It is possible to correlate per cent leaf injury with fluoride concentrations in leaves.[10,15,19] Correlations between per cent leaf injury, fluoride concentrations in leaves and atmospheric fluoride concentrations, however, may or may not occur.[2,10,18,19]

Gladiolus for use in biomonitoring plots must be grown for a while in a greenhouse or chamber where fluoride is not present (Table 33). Susceptible varieties—such as Snow Princess—and resistant varieties should be grown together to allow for comparisons.[19] Good cultural practices are essential in order to avoid plant responses that are confusing. Samples should be taken at least once a month for leaf injury evaluation and fluoride analyses.

The extent of foliar fluoride injury on gladiolus leaves has been

TABLE 32

EFFECTS OF FLUORIDE ON GLADIOLUS (*Gladiolus hortulanus* Bailey)

Corms	Reduction in size and weight
Flowers	Water-soaked tissue at margins of sepals and petals, drying and turning necrotic
Leaves	Light tan to ivory discoloration, with a dark brown margin, extending from tips downwards or along the margins. Varieties with lighter flower colors often have light-colored necrotic areas. An even front of necrotic tissue indicates continual chronic exposure to fluoride. A series of brown bands in the necrotic areas indicates successive separate exposures to fluoride. Each band represents a response to an exposure.

Adapted from references 3, 10, 11, 15, 18 and 20.

TABLE 33

PROCEDURES FOR USING GLADIOLUS (*Gladiolus hortulanus* Bailey) AS A BIOMONITOR FOR ATMOSPHERIC FLUORIDE

(1) Select superior corms (Nos 3 or 4) of a fluoride-sensitive variety such as Snow Princess and a fluoride-resistant variety such as Mansoer.
(2) Pot corms individually or in groups of between four and six in pots in fluoride-free greenhouse.
(3) When plants are 25 cm tall, or at the 4–5 leaf stage, transfer the plants to field plots.
(4) Observe good cultural practices. Provide water and plant nutrients. Control weeds, insects and diseases of microbial origin.
(5) Evaluate the plants at least at monthly intervals for per cent leaf injury (see Table 34).
(6) At each evaluation, save samples of necrotic and healthy leaf tissue for fluoride analyses.

Adapted from references 2, 16, 19 and 20.

TABLE 34

COMPARATIVE METHODS FOR EVALUATING FLUORIDE INJURY ON LEAVES OF GLADIOLUS (*Gladiolus hortulanus* Bailey)

Determine per cent leaf injury
(1) Visual estimation in the field.[19]
(2) Use planimeter to trace injured areas on leaves.[16]
(3) Photocopy leaves and use planimeter to trace photocopies.[20]

Determine leaf injury index[2]
Measure length of injured and healthy areas on all leaves on all plants.

$$\text{Injury index} = \frac{\text{Total length of injured area} \times 100}{\text{Total leaf area}}$$

estimated in a number of ways (Table 34). The simplest method is to make visual estimates of per cent of damaged leaf areas in the field and combine these with leaf analysis data for fluoride. Van Raay[19] did this with Snow Princess (susceptible) and Mansoer (resistant) gladiolus in plots in the south-west part of The Netherlands. A compensating planimeter can be used to trace and determine areas of fluoride-injured and healthy leaf tissues.[16, 20] This same technique can be used to determine injured areas on photocopies of leaves from monitoring plots.[20] By measuring the length of injured and healthy areas on all leaves from each gladiolus plant, used in plots near Spokane, Washington, USA, Adams *et al.*[2] were able to determine a numerical leaf injury index by multiplying the total length of injured area by 100 and dividing by total leaf area.

The office photocopier system of Vasiloff and Smith[20] can be used to quickly and permanently record fluoride injury on gladiolus leaves. These workers developed a reusable template (Table 35) that contained all the essential information about the leaf samples to be copied. Four middle-aged leaves were included with the template and a photocopy showing both was made. The photocopy became a permanent record. At a later date, a compensating planimeter was used to determine injured leaf areas shown

TABLE 35

SAMPLE REUSABLE TEMPLATE USED BY VASILOFF AND SMITH[20] WITH GLADIOLUS LEAVES (*Gladiolus hortulanus* Bailey) AND THE OFFICE PHOTOCOPIER SYSTEM FOR RECORDING FLUORIDE INJURY

Gladiolus Leaf Injury

Area ____________________

Collection Date ____________________

Plot No. ____________________

Plant No. ____________________

Leaf No. ____________________

Remarks ____________________

	Measured area (cm^2)	*Area as % of ELA*
ELA		
T & MN		
T & MC		
Other		
Total		

Code: ELA Exposed leaf area.
T & MN Terminal and marginal necrosis.
T & MC Terminal and marginal chlorosis.
Other Other types of injury.

TABLE 36

COMPARISON OF FLUORIDE CONTENT OF LEAVES AND PER CENT LEAF INJURY FOR GLADIOLUS (*Gladiolus hortulanus*, Snow Princess) FROM PLOTS AROUND A FERTILIZER PLANT IN ONTARIO

Time of growing season when samples were taken	*Distance and downwind direction from source*		
	3·4 km NE	*2 km E*	*5 km E*
July			
Fluoride content[a]	5·0 ppm	4·0 ppm	3·0 ppm
Leaf injury	4·6%	0·4%	0·2%
August			
Fluoride content	12·5 ppm	6·0 ppm	8·5 ppm
Leaf injury	8·8%	0·8%	0·7%
September			
Fluoride content	21·9 ppm	9·3 ppm	5·2 ppm
Leaf injury	15·7%	1·9%	1·3%

[a]Fluoride content in ppm, dry weight basis.
Adapted from Vasiloff and Smith.[20]

on the photocopy. This method was very useful where thousands of leaves had to be rapidly evaluated.

An example of the type of data that can be obtained with the photocopy method is given in Table 36. It shows that the fluoride content of gladiolus leaves decreased with downwind distance from the source, as did per cent leaf injury. The data also illustrate that fluoride accumulated in gladiolus plants as the growing season progressed.

2. Freesia

Freesias (*Freesia hybrida* Hort.) are a common horticultural plant in The Netherlands. Van Raay[19] used freesias to monitor fluoride around Rotterdam and near Pernis. Visual estimates of per cent damaged leaf area were correlated with fluoride analyses of leaves.

In plots near Pernis, Van Raay combined gladiolus and freesia with endive, clover, buckwheat and barley as biomonitors. He considered gladiolus and freesia to be more sensitive to fluoride than SO_2 and that clover and barley are more sensitive to SO_2 than fluoride. As most of the foliar injury symptoms occurred on endive, clover, buckwheat and barley, he concluded that SO_2 was probably the major air pollutant near Pernis.

IV. BIOMONITORING FLUORIDE WITH LOWER PLANTS

A. Lichens

In addition to taking up sulfur from aqueous solutions, lichens also take up fluoride. The same chemical methods used to analyse for fluoride in higher plants are also used to analyse for fluoride in lichen thalli.

Lichens respond to atmospheric fluoride in very much the same way that they respond to SO_2. The overall effect of fluoride on lichen thalli is chlorosis, followed by necrosis and thallus disintegration. Concentric zones of lichen distribution develop around fluoride sources (Table 37). Each zone has its own lichen flora.

Nash[13] determined the effects of fluoride from an isolated chemical factory in Pennsylvania on the transplanted terricolous lichens, *Cladonia cristatella* and *C. polycarpoides*, and on the saxicolous lichen, *Parmelia plittii.* All were killed by fluoride at a distance of 100 m from the factory.

The pattern of lichen distribution on the tops of wooden fence posts around an aluminum factory at Fort William, Scotland, was determined by Gilbert.[7] As he moved from a normal or a clean zone to near the factory, he found that fluoride concentrations of lichen thalli increased and the luxuriance of lichen cover on the tops of fence posts progressively became reduced until the post tops were bare. He established three lichen zones (Table 37) around the factory. Near the factory, most lichens were eliminated. In the transition zone away from the factory, he found declining lichen cover, suppression of fruiting and lack of luxuriance. In the normal or clean zone, lichens appeared abundant and normal.

The fluoride content of pine needles in Sweden[6] was shown to be related to the presence of the lichens *Pseudevernia furfuracea* and *Alectoria implexa.*

TABLE 37

EFFECT OF AIRBORNE FLUORIDE ON LICHENS ON THE TOPS OF WOODEN FENCE POSTS AROUND AN ALUMINUM FACTORY AT FORT WILLIAM, SCOTLAND, AS DETERMINED BY GILBERT[7]

Lichen zones	*Lichen responses*
Inner zone near the factory	Elimination
Transition zone away from factory	Declining cover, suppression of fruiting, lack of luxuriance
Normal zone	Luxuriance

Eriksson[6] found that both lichens were absent when the fluoride content of old pine needles reached 41 ppm or greater.

Given sensitive indicator plants, reliable fluoride analyses of plant tissues and air monitoring data for fluoride, it is possible to use both higher plants and lichens to biomonitor atmospheric hydrogen fluoride. The comparative advantages and disadvantages of using plants to monitor fluoride are given in Table 27.

REFERENCES

1. ADAMS, D. F. (1963). Recognition of the effects of fluorides on vegetation. *J. Air Poll. Contr. Assoc.* **13**: 360–2.
2. ADAMS, D. F., C. G. SHAW, R. M. GNAGY, R. K. KOPPE, D. J. MAYHEW and W. D. YERKES, JR. (1956). Relationship of atmospheric fluoride levels and injury indexes on gladiolus and Ponderosa pine. *J. Agric. Food Chem.* **4**: 64–66.
3. BREWER, R. F., F. B. GUILLEMET and F. H. SUTHERLAND. (1966). The effects of atmospheric fluoride on gladiolus growth, flowering, and corm production. *Proc. Am. Soc. Hort. Sci.*, **88**: 631–4.
4. COOKE, J. A., M. S. JOHNSON and A. W. DAVISON. (1976). Determination of fluoride in vegetation: A review of modern techniques. *Environ. Pollut.* **11**: 257–68.
5. DE ONG, E. R. (1946). Injury to apricot by fluoride deposit. *Phytopathology* **36**: 469–71.
6. ERIKSSON, O. (1966). Lavar och luftfororeningar i Sundsvallstrakten. Uppsala: Vaxtbiol. Inst.
7. GILBERT, L. L. (1973). The effect of airborne fluorides. *Air Pollution and Lichens.* (Ferry, B. W., Baddeley, M. S. and Hawksworth, D. L. (eds.)), Athlone Press London, pp. 176–91.
8. GILBERT, O. L. (1975). Effects of air pollution on landscape and land-use around Norwegian aluminum smelter. *Environ. Pollut.* **8**: 113–21.
9. HAWKSWORTH, D. L. and F. ROSE. (1976). *Lichens as Pollution Monitors.* Edward Arnold Pub., London. 59 pp.
10. JOHNSON, F., D. F. ALLMENDINGER, V. L. MILLER and C. J. GOULD. (1950). Leaf scorch of gladiolus caused by atmospheric fluoric effluents. *Phytopathology* **40**: 239–46.
11. LACASSE, N. L. and M. TRESHOW. (1976). *Diagnosing Vegetation Injury caused by Air Pollution,* US EPA Publication.
12. MCCUNE, D. C., A. E. HITCHCOCK and L. H. WEINSTEIN. (1966). Effects of mineral nutrition on the growth and sensitivity of gladiolus to hydrogen fluoride. *Contrib. Boyce Thompson Inst.*, **23**: 295–300.
13. NASH, T. H. (1971). Lichen sensitivity to hydrogen fluoride. *Torrey Bot. Clut Bul.* **98**: 103–6.
14. ROBAK, H. (1969). Aluminum plants and conifers in Norway. *Proc. First Eur. Congress on the Influence of Air Pollution on Plants and Animals, Wageningen,* pp. 27–31.

15. SPIERINGS, F. H. F. G. (1967). Chronic discoloration of leaf tips of gladiolus and its relation to the hydrogen fluoride content of the air and the fluorine content of the leaves. *Neth. J. Pl. Path.* **73**: 25–28.
16. TEMPLE, P. J., S. N. LINZON and M. L. SMITH. (1978). Fluorine and boron effects on vegetation in the vicinity of a fiberglass plant. *Water, Air and Soil Poll.*, **10**: 163–74.
17. TRESHOW, M., F. K. ANDERSON and F. HARNER. (1967). Responses of Douglas fir to elevated atmospheric fluorides. *For. Sci.*,**13**: 114–20.
18. TRESHOW, M. and M. R. PACK. (1970). Fluoride. In: *Recognition of Air Pollution Injury to Vegetation: A Pictorial Atlas.* (Jacobson, J. S. and A. C. Hill (eds.)), Air Poll. Contr. Assoc., Pittsburgh, pp. D1–D17.
19. VAN RAAY, A. (1969). The use of indicator plants to estimate air pollution by SO_2 and HF. *Proc. First Eur. Congress on the Influence of Air Pollution on Plants and Animals, Wageningen*, pp. 319–328.
20. VASILOFF, G. N. and M. L. SMITH. (1974). A photocopy technique to evaluate fluoride injury on gladiolus in Ontario, *Plant Disease Reptr.* **58**: 1091–4.

CHAPTER 7

Biomonitoring Heavy Metals and Dusts

I. INTRODUCTION

In addition to being exposed to potentially phytotoxic gases, such as O_3, SO_2 and HF, plants are also continuously exposed to airborne particles. These may consist of relatively inert substances, like alkaline dusts from limestone quarries and cement plants, or forms of one or more heavy metals, such as cadmium (Cd), copper (Cu), iron (Fe), lead (Pb), mercury (Hg), nickel (Ni) and zinc (Zn). Many of these particulates are blown off or washed away, while others remain on plant leaves, or may enter leaves via stomates and injuries.

Higher plants have been used as collectors of airborne dusts and heavy metals.[6,8,10,19,25] Higher plant leaves, however, are also likely to contain heavy metals obtained from soil via roots, as well as those obtained directly from atmospheric fall-out. Unless soil concentrations and rates of plant uptake are known, it can become difficult to interpret data on heavy metals obtained from plant leaf analyses alone.

Lower plants, especially mosses and lichens, have been used extensively to biomonitor heavy metals and occasionally dusts.[2,7–9,12,15–21,23,24] Unlike higher plants, they lack organized roots and vascular systems. All elements found in plants and plant parts are directly absorbed from air or by precipitation. This should make it possible to make good correlations between the heavy metal content of a moss or a lichen, ambient concentrations of the element and deposition rates of the element.

II. PHYSICAL AND CHEMICAL ANALYSES

A. Physical Analyses

The physical burden of inert dusts and particulate heavy metals on leaves and other plant parts can be determined rather easily. Surface particles can

TABLE 38
EFFECTS OF LIMESTONE DUST ON SECOND-YEAR NEEDLES OF HEMLOCK (*Tsuga canadensis* L.)

Locations	*Needle dry weights (g)*[a]	*Dust burden (mg)*
Dusty area		
Location 1	0·90	290
2	0·81	298
3	1·20	307
4	1·20	258
5	1·13	348
Clean area		
1	1·50	–
2	1·53	–
3	1·65	–
4	1·25	–
5	1·67	–

[a]Average of 500 needles per location, chosen at random.
Previously unpublished data: W. J. Manning.

be washed off leaves or known leaf areas. Careful evaporation of the washings will leave the dried particulates, which can then be weighed and also used for chemical analyses. Using this technique, it was possible to calculate the accumulated burden of limestone dust on hemlock needles (Table 38).

B. Chemical Analyses

A variety of standard methods can be used to determine the heavy metal content of particulates from plant surfaces and in plant tissues. The most common analytical method is atomic absorption spectrophotometry. Plasma emission spectrophotometry, X-ray fluorescence and neutron activation analysis are also used.

Vegetation samples must be thoroughly washed with EDTA: detergent mixes and rinsed with deionized water before processing for chemical analyses. This will remove casual dust and heavy metals. There is some debate about the efficiency of washing methods. Only a certain percentage of elements are removed by washing, e.g. only about 50% of Pb on leaves is removed by washing.[25]

Godzik *et al.*[5] attempted to determine whether Zn, Fe and Pb from zinc and lead smelters were taken up by red oak (*Querous robur* L.) and Scots pine (*Pinus sylvestris* L.) leaves. Chloroform and ultrasonic vibration were used to remove the wax and particulates from leaf surfaces prior to analysis, employing the X-ray fluorescence method. Removal of wax and particulates led to a decrease in Zn, Fe and Pb levels, even below the levels of the control leaves. It is possible that heavy metals coming in contact with leaves may be immobilized in the wax layer. As normal washing methods do not remove wax, the erroneous assumption may be made that the heavy metals were taken up by the leaf tissues. Scanning electron micrographs of leaf surfaces also suggested that stomates do not play a very active role in heavy metal uptake by leaves.

III. BIOMONITORING WITH HIGHER PLANTS

A. Heavy Metals

Vegetation along roadsides and in the vicinity of smelters has often been used to biomonitor heavy metal deposition, especially Pb.[25] Surface deposition can be easily determined and tissues analyzed for metal uptake. Little and Martin[11] collected leaves of elm (*Ulmus glabra*) and hawthorn (*Crataegus monogyna*) around a smelting complex and determined the deposition of Cd, Pb and Zn. They found that deposition of these heavy metals was greatly influenced by prevailing winds in relation to the smelter. The grass, *Festuca rubra*, was sampled along transects downwind of the Swansea urban-industrial complex in south-west Wales.[6] Concentrations of Cd, Ca, Ni, Pb and Zn were significantly greater along these transects than in the normal background concentrations of plant materials elsewhere. Everett *et al.*[3] surveyed the British Isles for deposition of airborne Pb. Private (*Ligustrum vulgare* L.) was used as a biomonitor. Privet leaves were collected from different locations and analyzed for Pb. As the leaves were not washed prior to Pb analysis, it was possible to determine and map locations with high and low atmospheric Pb levels.

B. Dusts

Most dusts are composed of inert, non-toxic particles. If they become too thick, the amount of sunlight available to leaves may be reduced and photosynthesis—and ultimately growth—may be reduced.[10] Cement and limestone dust may become thick and crust-like, especially on two- and three-year-old conifer needles, and reduce plant growth. Alkaline solutions from cement dust may cause injury to plant parts.[1] Alkaline dust deposit effects on spruce and fir branches are illustrated in Figs 34 and 35.

FIG. 34. Lime dust coating on spruce (*Picea*) branch (right). Note difference in size of needles when compared with non-dusted branch (left). (Courtesy: US EPA.)

FIG. 35. Cement dust on fir (*Abies*) branch. Note extensive loss of second-year needles. (Courtesy: US EPA.)

Manning[14] investigated the long-term effects of limestone dust deposition on hemlock (*Tsuga canadensis* L.). A 50 per cent reduction in the average length of new growth of terminals was found when results from trees growing in a heavy dust area were compared with those from trees in an area without dust (Table 39). Two-year-old needles from trees in the

TABLE 39
EFFECTS OF LIMESTONE DUST ON NEW GROWTH OF HEMLOCK (*Tsuga canadensis* L.)

Locations	*Average new growth of terminals in centimetres*[a] *Heavy dust*	*No dust*
1	3·6	8·1
2	3·2	7·6
3	5·4	7·2
4	4·5	8·6
5	5·6	12·1
Overall average	4·5	8·7

[a] Average of sixty terminals per location.
Adapted from Manning.[14]

dust area were chlorotic and needle dry weights were considerably reduced. Dust burdens for needles from each sampling location were also determined (Table 38).

IV. BIOMONITORING WITH LOWER PLANTS

A. Heavy Metals

Both mosses and lichens absorb heavy metals from air and by precipitation. A number of them have been used in field studies to biomonitor airborne heavy metals (Table 40). Changes in heavy metal concentrations of lichens and mosses usually directly reflect uptake from external sources.

In mosses, the uptake of heavy metal ions from solution is a process of ion exchange and chelation. This occurs in both living and dead tissues.[21] Metal uptake in lichens is a passive diffusion process by which different species selectively absorb metals. Metals are deposited either on the outer surface of the walls of the fungal symbiont or within fungal walls. Heavy metals are immobilized in the fungal partner and do not affect the algal symbiont until very high concentrations are accumulated.[8]

Smelters emit heavy metals and also SO_2. Separating SO_2 and metal effects on mosses or lichens can be difficult, unless their responses to both are known.[8] The lichen *Lecanora conizaeoides* is known to be a very efficient accumulator of heavy metals. As it is very tolerant to SO_2, it is well suited to be a good biological monitor of airborne heavy metals.[16]

TABLE 40

SOME MOSSES AND LICHENS COMMONLY USED IN FIELD STUDIES TO BIOMONITOR AIRBORNE HEAVY METALS

Mosses	*Lichens*
Dicranum polysetum	*Cladonia rungiferina*
Dicranum scoparium	*Hypogymnia physodes*
Hylocomium splendens	*Lecanora conizaeoides*
Hypnum cupressiforme	*Pseudoevernia furfuracea*
Pleurozium schreberi	*Usnea filipendula*
Pohlia nutans	
Sphagnum spp.	

Adapted from references 2, 4, 6, 7, 9, 16–19, 20 and 21.

1. Mosses

In terms of heavy metal uptake and accumulation, mosses seem to be more efficient than lichens. Mosses grow in large carpets on the ground and have more surface area to intercept particulate deposition and rain. Folkeson[4] found that the five mosses that he studied in Sweden (Table 41) accumulated greater concentrations of heavy metals than the four lichens used for comparison. Mosses were found to accumulate Hg by an order of one magnitude greater than grasses, or other vegetation, around Oak Ridge, Tennessee, USA.[9]

Ruhling and Tyler[18,20,21] have used the common woodland moss, *Hylocomium splendens*, to biomonitor heavy metal deposition in Scandinavia. *H. splendens* occurs in large carpets in most parts of Scandinavia. Each moss shoot has as many as seven segments, each representing one year's growth. The two youngest segments are green, the rest brown. By means of moss analyses, maps that show regional differences in atmospheric deposition of Cd, Cu, Fe, Hg, Ni, Pb and Zn have been prepared for Finland, Norway and Sweden. Pb concentrations increased with precipitation and decreased with distance from roads and cities.[18]

Both *Hylocomium splendens* and *Pleurozium schreberi* were used by Grodzinska[7] to determine heavy metals in twelve Polish national parks. Increased heavy metal contents of the mosses related well to locations near industrial centers. A strong correlation was found between the annual sum of precipitation and heavy metal concentrations in the mosses. Grodzinska concluded that mosses were very precise and useful bioindicators of heavy metals.

TABLE 41

MOSSES AND LICHENS USED TO BIOMONITOR HEAVY METALS AROUND A BRASS FOUNDRY IN SOUTHEAST SWEDEN

Plants	*Substrate*	*Plant parts used for metal analyses*
Mosses		
Dicranum polysetum	Ground	Top of shoot
Hylocomium splendens	Ground	Three youngest segments
Hypnum cupressiforme	Ground	Green and light-brown parts
Pleurozium schreberi	Ground	Green part
Pohlia nutans	Rocks or stones	Entire shoot
Lichens		
Cladonia rungiferina	Organic matter on rocks	Light-colored part
Hypogymnia physodes	*Pinus* trunks	Whole thallus
Pseudoevernia furfuracea	*Pinus* trunks	Whole thallus
Usnea filipendula	*Pinus* and *Picea* trunks	Whole thallus

Adapted from Folkeson.[4]

The common moss *Hypnum cupressiforme* is the most widely used moss in biomonitoring schemes for heavy metals.[2,4,6,18,19,23] Natural areas of *H. cupressiforme* have been sampled along transects for heavy metal accumulation.[23] Goodman and Roberts[6] developed a transplant technique that enabled them to use *H. cupressiforme* at measured distances from heavy metal sources. *H. cupressiforme*, from areas without heavy metal pollution, was placed in nylon mesh bags and transplanted to study areas. The bags were retrieved at weekly or monthly intervals and the moss tissues analyzed for Cd, Cu, Ni, Pb and Zn. Concentrations of these metals greater than normal were found downwind of the Swansea urban-industrial complex in south-west Wales. The *Hypnum* moss bag technique can be used to locate sources of airborne heavy metals and to determine long-term heavy metal burdens.

Little and Martin[12] used the nylon mesh bag technique with sphagnum moss (*Sphagnum* spp.) to monitor airborne Cd, Pb and Zn around a smelting complex at Avonmouth, near Bristol, Great Britain over a period of several months. By analyzing *Sphagnum* from 47 sampling sites, they were able to make computer maps showing Cd, Pb and Zn levels and

TABLE 42

COMPARATIVE MERCURY CONTENT OF THE MOSSES *Dicranum scoparium* AND *Polytrichum commune* FROM THREE SITES IN EAST TENNESSEE, USA, AT INCREASING DISTANCES FROM A FLY ASH SOURCE

Moss	*Mercury content of mosses (ppm) at sites*[b]		
	Oak Ridge (Near source)	*Cade's Cove (Intermediate)*	*Great Smoky Mountains (Farthest away)*
Dicranum scoparium[a]	1·13	0·118	0·066
Polytrichum commune	—	0·092	0·045

[a] *D. scoparium* forms dense low mats. *P. commune* grows more upright in less dense stands and, as a result, may entrap fewer fly ash particles.
[b] Average values.
Adapted from: Huckabee.[9]

patterns of distribution. These maps reflected the effects of prevailing winds, localized climatic conditions and topography on heavy metal emission and distribution.

Several mosses have been used to biomonitor airborne Hg.[9,24] Huckabee[9] studied the uptake of Hg from fly ash in the mosses *Dicranum scoparium* and *Polytrichum commune* at varying distances from smokestacks at Oak Ridge, Tennessee, USA (Table 42). He found that Hg concentration in these mosses was a function of distance from the smokestacks. Average concentrations in polluted areas ranged from 1·13 ppm to 0·066 ppm in remote non-polluted areas. The moss *D. scoparium* intercepted and accumulated more Hg than did *P. commune*. *Dicranum* forms dense low mats, while *Polytrichum* grows more upright, in less dense stands, and may intercept fewer fallout particles.

Rasmussen[17] used bryophytes on trees to determine heavy metal concentrations in rural areas of Denmark. He used the epiphytic bryophytes *Homalothecium sericeum, Hypnum cupressiforme, Isothecium myosuroides, I. myurum* and *Neckera complanata* as biomonitors. Heavy metal concentrations in these bryophytes were compared with that in dried herbarium specimens of the same bryophytes collected from the same trees in 1951. Rasmussen found one full order of magnitude increase in heavy metal concentrations in the last 25 years.

2. Mosses and lichens compared

It is known that moss and lichen species take up different heavy metals at different rates. In view of this, attempts to determine heavy metal burdens,

TABLE 43

HEAVY METAL CONCENTRATIONS (AVERAGE IN μg/g TOTAL DRY WEIGHT AND ± 95% CONFIDENCE LIMITS) IN MOSSES AND LICHENS SAMPLED AROUND A BRASS FOUNDRY IN SOUTHEAST SWEDEN

Plants	*Heavy metal concentrations in μg/g and 95% confidence limits*					
	Cd	*Cu*	*Fe*	*Ni*	*Pb*	*Zn*
Mosses						
Dicranum	0·9 ± 0·08	32·0 ± 3·7	574 ± 41	2·9 ± 0·2	31·8 ± 1·6	179 ± 13
Hylocomium	1·3 ± 0·1	39·0 ± 4·2	647 ± 47	4·7 ± 0·3	46·1 ± 3·0	159 ± 13
Hypnum	1·4 ± 0·1	36·2 ± 5·0	1700 ± 146	7·0 ± 0·5	106 ± 7·3	208 ± 18
Pleurozium	0·8 ± 0·08	29·1 ± 2·9	591 ± 57	2·8 ± 0·2	35·5 ± 2·6	166 ± 11
Pohlia	2·0 ± 0·1	41·5 ± 8·4	1270 ± 128	5·0 ± 0·2	68·7 ± 5·8	237 ± 24
Lichens						
Cladonia	0·5 ± 0·06	14·5 ± 2·3	442 ± 47	1·5 ± 0·1	22·8 ± 1·5	102 ± 10
Hypogymnia	1·1 ± 0·06	28·2 ± 4·0	832 ± 55	2·6 ± 0·1	22·6 ± 1·2	232 ± 23
Pseudoevernia	0·6 ± 0·03	35·0 ± 3·8	926 ± 67	2·8 ± 0·1	37·3 ± 2·1	237 ± 21
Usnea	0·6 ± 0·04	22·4 ± 2·7	614 ± 68	2·6 ± 0·2	27·0 ± 1·4	182 ± 18

Adapted from Folkeson.[4]

with only one or two species of moss or lichen, may not be successful, as a true picture of heavy metal deposition and uptake may not be obtained. Folkeson[4] addressed this problem by conducting an extensive study of the comparative heavy metal uptake of five mosses and four lichens near a brass foundry in southeast Sweden (Table 41). His objective was to attempt to equate heavy metal concentrations in each species with all other species used. By developing and using calibration factors, he attempted to develop a method whereby the estimated heavy metal content of a species, not present at a sampling site, could be estimated from heavy metal concentrations of species that were present at a sampling site.

Moss and lichen samples were obtained from 57 sampling sites in coniferous woodlands around the foundry. The mosses and lichens studied, collection substrates and plant parts used for metal analyses are given in Table 41.

Plant samples were analyzed by flame atomic absorption spectrophotometry for Cd, Cu, Fe, Ni, Pb and Zn. Average concentrations of each of these metals for each moss or lichen, together with $\pm$ 95% confidence limits for each value, are given in Table 43.

Using the data in Table 43, Folkeson was able to develop calibration

TABLE 44

CALIBRATION FACTORS (MEANS) FOR ESTIMATING TOTAL HEAVY METAL CONCENTRATIONS IN THE MOSS *Hylocomium splendens* FROM CONCENTRATIONS MEASURED IN ANY OF THE OTHER EIGHT MOSSES OR LICHENS

	Calibration factors (means) for elements					
Plants	*Cd*	*Cu*	*Fe*	*Ni*	*Pb*	*Zn*
Mosses						
Dicranum	1·5	1·25	1·14	1·6	1·46	0·90
Hypnum	0·9	1·11	0·39	0·68	0·44	0·78
Pleurozium	1·7	1·37	1·11	1·7	1·31	0·97
Pohlia	0·7	1·00	0·52	0·94	0·68	0·68
Lichens						
Cladonia	2·7	2·81	1·49	3·2	2·04	1·59
Hypogymnia	1·2	1·43	0·78	1·8	2·05	0·70
Pseudoevernia	2·2	1·14	0·71	1·7	1·24	0·68
Usnea	2·2	1·79	1·08	1·8	1·72	0·89

Adapted from Folkeson.[4]
Based on data in Table 43. See also Table 45.

TABLE 45

CALCULATION OF THE CALIBRATION FACTOR FOR ESTIMATING Fe CONCENTRATION IN *Hylocomium* FROM VALUES MEASURED IN *Dicranum*

The interspecies calibration method

(1) Consult Table 43.
Mean value for Fe in *Hylocomium* is 647 ± 47
Mean value for Fe in *Dicranum* is 574 ± 41

(2) According to the 95% confidence limits, the quotient between the means ranges between:

$$0{\cdot}976 \text{ (minimum value) } \frac{(647-47)}{(574-41)} = (0{\cdot}976)$$

and

$$1{\cdot}30 \text{ (maximum value) } \frac{(647+47)}{(574-41)} = (1{\cdot}30)$$

The mean of this interval is 1·14 (see Table 44).

(3) The calibration factor for estimating Fe concentrations in *Hylocomium* from values measured in *Dicranum* is thus 1·14.

(4) *Example* 150 μg/g dry weight of Fe in *Dicranum* is estimated as corresponding to 171 μg/g in *Hylocomium*.
($1{\cdot}14 \times 150\ \mu g/g = 171\ \mu g/g$).

Adapted from Folkeson.[4]

factors for all the elements and all the mosses and lichens. Calibration factors were determined for estimating total heavy metal concentrations in the moss *Hylocomium splendens* from concentrations measured in the other eight bryophytes (Table 44). An example of how calibration factors were determined is given in Table 45. The example given here is for estimating Fe concentrations in *H. splendens* from values measured in *D. polysetum*. An example of how the calibration factor can be used to equate Fe concentrations in *H. splendens* with that in *D. polysetum* is also given in Table 45.

Folkeson found considerable variability and selectivity in heavy metal uptake by the mosses and lichens he studied. Mosses were better accumulators of heavy metals than lichens. *Hypnum* accumulated the largest amounts of Fe and was very high in Ni and Pb. *Cladonia* accumulated the least amount of any element. *Pohlia* accumulated the most Cd, while *Dicranum* and *Pleurozium* had the lowest Cd, Ni and Pb.

3. Lichens

Pilegaard[16] found the crustose lichen *Lecanora conizaeoides* useful in monitoring airborne heavy metals around a foundry and a steelworks at Frederiksvaerk, Denmark. Both SO_2 and heavy metals were being emitted. SO_2 distribution was determined by mapping the occurrence and frequency of lichen species with known SO_2 sensitivities. *L. conizaeoides* is resistant to SO_2 and could be used as a specific biomonitor for heavy metals.

4. Spanish moss

Spanish moss (*Tillandsia useneoides*) is an epiphytic member of the bromeliad or pineapple family. It forms long hanging tufts of thread-like leaves on tree branches in warm tropical areas, especially the Gulf Coast area of the USA. Martinez *et al.*[15] collected samples of this false 'moss' and determined that it was a very efficient collector of Pb.

5. Puffball fungus

The 'gem-studded' puffball fungus, *Lycoperdon perlatum*, produces large sporophores in the autumn, especially in wooded areas. McCreight and Schroeder[13] collected *L. perlatum* sporophores at a roadside rest area in Connecticut adjacent to a major highway. They found that sporophores of *L. perlatum* were effective accumulators of Cd and Pb. Ni was only found in one sample.

B. Dusts

1. Lichens

Heavy dust deposition from limestone quarries and cement plants may result in white films or crusts on lichens and their substrates. Under these conditions, corticolous lichens are destroyed. Light to moderate alkaline dust deposition neutralizes acid substrates, such as bark. Species of the lichens *Xanthorion* and *Physcia*, normally confined to basic tree bark and stones, may begin to appear on trees, whose normally acid bark has been neutralized, with a resulting increase in pH. The presence of *Xanthorion* and *Physcia* on tree trunks where they normally would not be found can be used to determine the intermediate or transition zone between an area of high dust deposition, where lichens have been eliminated, and a clean or non-dusty area, where normal lichens prevail.[8]

2. Plant disease incidence

Alkaline dust deposition can also affect the susceptibility of certain plants to leafspot diseases caused by fungi. Schonbeck[22] found that cement kiln dust in Germany increased the incidence of leafspot of sugar beet, caused by the fungus *Cercospora beticola*. Fungal leafspot incidence on native vegetation, with and without a limestone dust burden, was determined by Manning.[14] He found that black rot spots on wild grape leaves, caused by the fungus *Guignardia bidwelli*, and anthracnose spots on

TABLE 46
EFFECT OF LIMESTONE DUST ON FUNGAL LEAFSPOT INCIDENCE ON SASSAFRAS AND WILD GRAPE

		Average number of leaf spots[a]	
Plant	*Fungus*	*Moderate dust*	*No dust*
Sassafras (*Sassafras albidum* Nutt.)	*Gloeosporium* sp.	6–7	0–1
Wild grape (*Vitis labrusca* L.)	*Guignardia bidwelli*	2–3	0–1

[a] Average of 40 leaves in five locations per area.
Adapted from Manning.[14]

TABLE 47
COMPARATIVE ADVANTAGES AND DISADVANTAGES OF USING PLANTS TO BIOMONITOR AIRBORNE HEAVY METALS AND DUSTS

Advantages

(1) Plants are excellent receptors and collectors of airborne heavy metals and dusts.
(2) Dusts and heavy metals, on and in plant parts, can be physically and chemically analyzed.
(3) Mosses and lichens selectively and directly take up heavy metals from air and rain.
(4) Plants can be used to determine the location of heavy metal and dust sources and patterns of distribution and deposition.

Disadvantages

(1) Higher plants also take up heavy metals from soil as well as air, making results difficult to interpret.
(2) Heavy metal content of higher plant leaves may reflect only what is trapped in the surface wax layer.
(3) Mosses and lichens are also sensitive to SO_2, HF and other pollutants.
(4) As mosses and lichens vary in their responses to heavy metals, a number of species are necessary for accurate biomonitoring.

Adapted from references 3–6, 8, 9, 11, 12, 16, 23 and 25.

sassafras leaves, caused by the fungus *Colletotrichum*, were more extensive on leaves with a dust burden than on those without one (Table 46).

Plants can be used successfully to biomonitor airborne heavy metals and dusts. Mosses seem to be the most widely used and successful biomonitoring plants. The comparative advantages and disadvantages of using plants to biomonitor heavy metals and dusts are summarized in Table 47.

REFERENCES

1. DARLEY, E. F. (1966). Studies on the effect of cement-kiln dust on vegetation. *J. Air Poll. Contr. Assoc*, **16**: 145–50.
2. ELLISON, G., J. NEWHAM, M. J. PINCHIN and I. THOMPSON. (1976). Heavy metal content of mosses in the region of Consett (North East England). *Environ. Pollut.* **11**: 167–74.
3. EVERETT, J. L., C. L. DAY and D. REYNOLDS. (1967). Comparative survey of lead at selected sites in the British Isles in relation to air pollution. *Food Cosmet. Toxicol.*, **5**: 29–35.
4. FOLKESON, L. (1979). Interspecies calibration of heavy-metal concentrations in nine mosses and lichens: Applicability to deposition measurements. *Water, Air and Soil Poll.*, **11**: 253–60.
5. GODZIK, S., T. FLORKOWSKI, S. PIOREK and M. A. SASSEN. (1979). An attempt to determine the tissue contamination of *Quercus robur* L. and *Pinus sylvestris* L. foliage by particulates from zinc and lead smelters. *Environ. Pollut.*, **18**: 97–106.
6. GOODMAN, G. T. and T. M. ROBERTS. (1971). Plants and soils as indicators of metals in the air. *Nature (London)*, **231**: 287–92.
7. GRODZINSKA, K. (1978). Mosses as bioindicators of heavy metal pollution in Polish national parks. *Water, Air and Soil Poll.*, **9**: 83–97.
8. HAWKSWORTH, D. L. and F. ROSE. (1976). *Lichens as PollutionMonitors.* Edward Arnold Publishers, London, 59 pp.
9. HUCKABEE, J. W. (1973). Mosses: Sensitive indicators of airborne mercury pollution. *Atmos. Environ.*, **7**: 749–54.
10. LACASSE, N. L. and M. TRESHOW. (1976). *Diagnosing Vegetation Injury Caused by Air Pollution.* US EPA Publication.
11. LITTLE, P. and M. H. MARTIN. (1972). A survey of zinc, lead and cadmium in soil and natural vegetation around a smelting complex. *Environ. Pollut.*, **3**: 241–54.
12. LITTLE, P. and M. H. MARTIN. (1974). Biological monitoring of heavy metal contamination. *Environ. Pollut.*, **6**: 1–19.
13. MCCREIGHT, J. D. and D. B. SCHROEDER. (1977). Cadmium, lead, and nickel content of *Lycoperdon perlatum* Pers. in a roadside environment. *Environ. Pollut.*, **13**: 265–8.
14. MANNING, W. J. (1971). Effects of limestone dust on leaf condition, foliar disease incidence, and leaf surface microflora of native plants. *Environ. Pollut.*, **2**: 69–76.

15. MARTINEZ, J. D., M. NATHANY and V. DHARMARAJAN. (1971). Spanish moss, a sensor for lead. *Nature (London)*, **233**: 564–5.
16. PILEGAARD, K. (1978). Airborne metals and SO_2 monitored by epiphytic lichens in an industrial area. *Environ. Pollut.*, **17**: 81–92.
17. RASMUSSEN, L. (1977). Epiphytic bryophytes as indicators of the changes in the background levels of airborne metals from 1951–1975. *Environ. Pollut.*, **14**: 37–61.
18. RUHLING, A. and G. TYLER. (1968). An ecological approach to the lead problem. *Bot. Notiser.*, **121**: 321–42.
19. RUHLING, A. and G. TYLER. (1969). Ecology of heavy metals—A regional and historical study. *Bot. Notiser.* **122**: 248–59.
20. RUHLING, A., and G. TYLER. (1970). Sorption and retention of heavy metals in the woodland moss *Hylocomium splendens*. (Hedw.). Br. et. Sch., *Oikos.*, **21**: 92–97.
21. RUHLING, A. and G. TYLER. (1973). Heavy metal deposition in Scandinavia. *Water, Air and Soil Poll.* **2**: 445–55.
22. SCHONBECK. H. (1960). Beobachtung zur Frages des Einflusses von industriellen Immisionen auf die Krankbereitschaft der Pflanze. *Ber Landesanstalt fur Bodennutzungsschutz*, **1**: 89–98.
23. WALLIN, T. (1976). Deposition of airborne mercury from six Swedish chloralkali plants surveyed by moss analysis. *Environ. Pollut.* **10**: 101–14.
24. YEAPLE, D. S. (1972). Mercury in bryophytes (moss). *Nature (London)*, **235**: 229–30.
25. ZIMDAHL, R. L. (1976). Entry and movement in vegetation of lead derived from air and soil sources. *J. Air Poll. Control Assoc.*, **26**: 655–60.

CHAPTER 8

Biomonitoring Ethylene

I. INTRODUCTION

Ethylene (C_2H_4) is becoming a very common air pollutant. It has many natural sources and is associated with manufacturing and automobile exhaust emissions.[1,2,5,6] Ethylene also accumulates when plant material is stored in poorly ventilated areas.[7] The general effect of ethylene is a reduction in vegetative growth and flower and fruit development and an acceleration of normal ageing of plant tissues.[2,5]Ethylene may also interact with other air pollutants and result in injury that is additive in nature.[5]

Ethylene affects plants at very low concentrations. Whilst most air pollutants are measured in parts per million (ppm), ethylene is measured in parts per billion (ppb). The threshold for many plants, under experimental conditions, is 10 ppb (Table 48). Some plants, like African marigold and orchid, react to ethylene at concentrations below 10 ppb.[3–5]

Little systematic monitoring of ambient concentrations of ethylene has been done. Abeles and Heggestad[2] mapped the metropolitan Washington, DC, USA, area in 1972. Air samples were taken at a number of locations radiating out from the Capitol building in central Washington. Ethylene concentrations were found to be highest in central Washington and to decrease with distance to the Virginia and Maryland suburbs (Table 49).

The effects of ethylene on plants under experimental (laboratory) conditions are quite well-known[1] and several lists of sensitive plants and their responses have been compiled.[2,4–6] Responses of plants to ethylene under natural or ambient conditions, however, have not been extensively studied. Biomonitoring schemes for ethylene have not been developed. In view of the increasing importance of ethylene as an air pollutant, several plants suitable as possible biomonitors are discussed here.

TABLE 48

CALCULATED DOSE–RESPONSE VALUES FOR ETHYLENE

Responses in plants	*Ethylene concentration in ppb*
Threshold	10
Half maximum	100–500
Saturation	1000–10 000

Adapted from Abeles and Heggestad.[2]

TABLE 49

TYPICAL AMBIENT CONCENTRATIONS OF ETHYLENE AT SEVERAL LOCATIONS IN THE METROPOLITAN WASHINGTON, DC, USA, AREA IN 1972

Locations	*Average concentration of ethylene in ppb*
Central Washington	600–700
North-west at National Zoo	320
Outer Beltway in Virginia	150
Beltsville, Maryland (USDA)	39

Adapted from Abeles and Heggestad.[2]

II. POSSIBLE BIOMONITORS

A. Orchid

Davidson[3] made a thorough investigation of the response of orchid flowers (*Cattleya* spp.) to ethylene, under both experimental and natural conditions. He found that young orchid flowers, with sepals begining to split, were very sensitive to low concentrations of ethylene. Mature flowers, however, were usually not affected. Young sepals turned brown, beginning at the tips, and later became necrotic and dry (Fig. 36). This response was termed 'dry sepal' and it is the best known and most reliable symptom of acute plant injury caused by ethylene.[6] Under laboratory conditions, dry sepal occurred on young flowers when exposed to 2 ppb ethylene for 24 h or 10 ppb for 8 h.

Davidson noted that periodic reports of dry sepal injury on orchid flowers in New Jersey and New York coincided with calm, humid, cloudy

FIG. 36. Dry sepal of orchid (*Cattleya* sp.), caused by ethylene (C_2H_4). (Courtesy: USDA.)

weather during autumn and early winter months, when barometric pressure readings were below normal. He reported an episode in late November, 1947 during which dry sepal occurred on orchid flowers within a 50-mile radius of New York City. A similar incidence of widespread dry sepal on orchid flowers occurred in the metropolitan Boston area during an inversion period in late November of 1966 (W. A. Feder, pers. comm.)

Young *Cattleya* orchid flowers may have potential as very sensitive biomonitors of ambient ethylene. Their very exacting cultural requirements, however, would make them difficult to use on a large scale or in areas where greenhouses are not available.

B. Carnation

It has long been known that florists' carnations (*Dianthus caryophyllus* L.) respond to ethylene by prematurely ageing. Petals wilt and close and the flowers appear to be 'sleepy'. This is a normal ageing response in carnations and other cut flowers. Ethylene accelerates the response and reduces the length of time that carnations can be used as cut flowers. The time interval between cutting the carnation flowers and the first appearance of petal wilt can be determined without difficulty. This period is also known as the 'vase life' of cut carnations.

TABLE 50

USEFUL LIFE IN DAYS OF CUT CARNATIONS (*Dianthus caryophyllus* L. 'Red Sim') EXPOSED TO 50 PPB ETHYLENE ALONE AND IN COMBINATION WITH 2·27% CO_2

Days exposed to air regimes	*Useful life in days of cut carnations*[a]		
	No ethylene[b]	*Air plus 50 ppb ethylene*	*Air plus ethylene plus 2·27% CO_2*
2	8·6	5·8	8·9
4	8·0	5·3	9·7

[a]Cut carnations placed in water containing 0·003% silver nitrate as a bactericide.
[b]Cut flowers placed in a moving stream of air.
Adapted from Smith and Parker.[8]

Cut carnations might be useful as biomonitors of atmospheric ethylene. Smith and Parker[8] and Smith *et al.*[9] developed a system for exposing cut carnations to streams of air containing no ethylene or measured amounts of ethylene. Using vase life or useful life as a criterion, the effects of ethylene on cut carnations could be determined. 50 ppb ethylene for 2 days decreased the useful life of cut Red Sim carnations by 2·8 days (Table 50). Carbon dioxide (CO_2) concentrations as low as 2·2% were found to inhibit ethylene effects on carnations.

An observation that CO_2 concentrations of less than 1% may occur in urban areas during fog[8] suggests that this may be one reason why ethylene injury occurs periodically on carnations and other flowers in greenhouses during periods when the air is stagnant.

C. Cucumber

Abeles and Heggestad[2] investigated the effects of ethylene on a wide range of plants under laboratory conditions at the USDA, Beltsville, Maryland, USA. They were able to evaluate plants grown in an environment from which most ethylene had been excluded and at known concentrations of ethylene.

Their experimental regime might be adapted to an air monitoring system using plants as biomonitors. Plants were grown in chambers where incoming air was passed through a Purafil filter ($KMnO_4$ absorbed on alumina pellets—H. E. Burroughs and Assoc., 3550 Broad St., Chamblee, Georgia, USA.). This filter was approximately 75% efficient in removing

TABLE 51

RESPONSE OF CUCUMBER (*Cucumis sativus* L.), TOMATO (*Lycopersicon esculentum* Mill.) AND HONEY LOCUST (*Gleditsia triacanthos* L.) TO ETHYLENE FUMIGATION UNDER EXPERIMENTAL CONDITIONS

Plant	*Ethylene concentration (ppb)*	*Response*
Cucumber	25	Inhibition of leaf expansion
	50	Reduction in flowering Female flowers increased (3 : 1 ratio)
	100	All female flowers
Tomato	25	Beginning of epinasty
Honey locust	75	Yellowing and abscission of leaflets Hypertrophied lenticels

Adapted from Abeles and Heggestad.[2]

ethylene (and most SO_2, O_3, H_2S and NH_3 as well). Plants grown in such chambers could be moved to urban areas and their growth and development compared with similar plants kept in chambers with Purafil filters.

Cucumber (*Cucumis sativus* L.) was a very sensitive plant under experimental conditions (Table 51). At 25 ppb ethylene, leaf expansion was inhibited—a response easily quantified. At 50 ppb, the number of female flowers increased and became dominant at 100 ppb. Cucumber might make a good biomonitor for ambient ethylene as it has two good responses that can easily be measured. Concentrations required for these responses were not unusual in ambient air in the Washington area (Table 49).

D. Tomato and Marigold

Tomato (*Lycopersicon esculentum* Mill.) and African marigold (*Tagetes erecta* L.) respond to low concentrations of ethylene in a classic way called epinasty. Petioles bend downwards and leaves droop visibly. Marigold has been reported to do this under experimental conditions at ethylene concentrations as low as 25 ppb (Table 51).

Using the previously described system of Abeles and Heggestad,[2] epinasty in tomato and marigold might be a useful biomonitoring symptom for atmospheric ethylene. Other causes of epinasty in both plants are possible, including ethylene released during the development of vascular wilt diseases, caused by fungi such as *Fusarium*.

E. Honey Locust

Abeles and Heggestad[2] observed two obvious responses in one-year-old honey locust trees (*Gleditsia triacanthos* L.) exposed to 75 ppb ethylene for 30 days (Table 51). Leaflets turned yellow and abscissed while stems developed hypertrophied lenticels (intumescences). Potted honey locust trees might make useful biomonitors of ambient ethylene. As woody plants they might provide more cumulative information than do annual plants.

Many other environmental factors can cause premature yellowing and leaf fall in trees such as honey locust. Drought, insects, heat, etc., could be complicating and confusing factors. Hypertrophied lenticels, however, are a classic ethylene response.[1]

REFERENCES

1. Abeles, F. B. (1973), *Ethylene in Plant Biology*. Academic Press, N. Y. 302 pp.
2. Abeles, F. B. and H. E. Heggestad. (1973). Ethylene: An urban air pollutant. *J. Air Poll. Contr. Assoc.*, **23**: 517–21.
3. Davidson, O. W. (1949). Effects of ethylene on orchid flowers. *Proc. Am. Soc. Hort. Sci.*, **53**: 440–6.
4. Heck, W. W., R. H. Davies and E. G. Pires (1962). *Effect of ethylene on horticultural and agronomic plants*. Texas Agr. Expt. Sta. Rept. MP-613.
5. Heck, W. W., R. H. Davies and I. J. Hindawi, (1970). Other phytotoxic pollutants. In: *Recognition of Air Pollution Injury to Vegetation*. (Jacobson, J. S. and A. C. Hill (eds.)), Air Poll. Contr. Assoc., Pittsburgh, Pa. pp. F1 to F24.
6. Lacasse, N. L. and M. Treshow. (1976). *Diagnosing Vegetation Injury Caused by Air Pollution*. US EPA Publication.
7. Smith, W. H., D. F. Meigh and J. C. Parker. (1964). Effect of damage and fungal infection on the production of ethylene by carnations. *Nature (London)*, **204**: 92–93.
8. Smith, W. H. and J. C. Parker. (1966). Prevention of ethylene injury to carnations by low concentrations of carbon dioxide. *Nature (London)*, **211**: 100–11.
9. Smith, W. H., J. C. Parker and W. W. Freeman. (1966). Exposure of cut flowers to ethylene in the presence and absence of carbon dioxide. *Nature (London)*, **211**: 99–100.

CHAPTER 9

Biomonitoring Air Pollutants with Plants: Future Directions

I. INTRODUCTION

We have seen in previous chapters that a number of woody and herbaceous higher plants, as well as mosses, lichens and fungi, can be used as indicators and biomonitors of gaseous and particulate air pollutants. Our purpose here will be to consider possible future needs and directions in biomonitoring air pollutants with plants.

II. NEED FOR MORE AND BETTER BIOMONITORS

A. Better Genetic Definition in Biomonitors

Many biomonitoring plants respond to more than one pollutant and interpreting plant responses can be difficult. In other plants, responses to pollutants may not be consistent or symptoms may not be very clear. There is a need to develop plants with similar stable phenotypes, but with genotypes differing only in sensitivity to a specific pollutant (isolines), for use in field biomonitoring schemes. The genetic markers must be specific and obvious so that resistant and susceptible isolines can be readily compared. Once genetically defined isolines are available, it is also essential to determine how environmental factors affect their response to an appropriate pollutant. These improvements will help to eliminate much of the uncertainty in interpreting the responses of plant biomonitors in the field.

B. Herbaceous versus Perennial Biomonitors

Most biomonitoring plants are herbaceous or are plants that must be planted out at monitoring sites each season. Several of them (e.g. Bel-W3 tobacco) require special care in order to respond to pollutants in characteristic fashion. There is a need to identify and utilize perennial biomonitor plants with or in place of herbaceous biomonitors. This would allow the establishment of long-term studies without interruption and with fewer interfering variables.

C. Central Sources of Biomonitors

It is often difficult for investigators to obtain the same plant materials used by someone else to biomonitor for a pollutant. There is a need to be able to readily obtain quantities of known pollutant-sensitive and pollutant-resistant plant materials. Several central facilities need to be developed to identify, evaluate, maintain, propagate and distribute plants that can be used as biomonitors for specific pollutants. These facilities would serve the same purpose as germ plasm banks do for geneticists and plant breeders.

D. Specific Pollutants

Examples of specific pollutants and future needs and directions for biomonitoring are given below.

1. Oxidants

Bel-W3 tobacco and Tempo garden bean can be used very effectively in biomonitoring schemes for O_3. Both, however, are annuals and must be replanted every year. Bel-W3 tobacco requires shading and special care to respond in characteristic fashion. White pine can be used as a perennial biomonitor, but SO_2 also contributes to the chlorotic dwarf syndrome. Grape vines, especially the cultivar 'Ives', may prove to be good perennial biomonitors of O_3 alone. One of the present authors (W. J. Manning) is currently developing a biomonitoring scheme for O_3 with grapes. Romaine lettuce and Pinto bean have been used to detect PAN, but neither seems to be especially good as a biomonitor. Perennial biomonitors for PAN need to be discovered and monitoring schemes developed. NO_2, and other oxides of nitrogen, do not appear to be serious air pollutants, except in glasshouses. Tomato and lettuce appear to be useful as biomonitors of NO_x, under confined conditions.

2. Sulfur dioxide

Conifers, certain deciduous trees, ferns and lichens seem to be excellent biomonitors of SO_2. There is also a need to identify more biomonitors, such as in lichens, where sensitivities to SO_2, HF and heavy metals are known, so that the effects of each can be determined separately.

3. Hydrogen fluoride
Apricot and peach, conifers, gladiolus, freesia and sometimes lichens, are commonly used to biomonitor HF. Reasonably good correlations can be made between HF in plant tissues and probable ambient HF concentrations.

4. Ethylene
Ethylene is becoming a pervasive air pollutant with pronounced effects on plants at extremely low concentrations. Orchids have been well documented as sensitive biomonitors, but their widespread use is not practical. Determining epinastic responses in African marigolds or tomatoes, under ambient conditions, is difficult. A good biomonitoring scheme for ethylene is needed that employs sensitive plants with easily determined responses under field conditions.

5. Heavy metals
Mosses seem to be the best biomonitors of heavy metals. Considerable work is needed to follow up on the observation that different mosses take up different metals at different rates. Mosses that accumulate specific metals, at known rates, need to be identified and used to further refine biomonitoring schemes.

III. PLANT BIOMONITORS AND AIR QUALITY STANDARDS

When sensitive plant biomonitors are used in well co-ordinated schemes with mechanical air monitoring devices, it is possible to obtain a good deal of information about the response of the biomonitor plants at known pollutant concentrations. Whilst this information is important and useful in itself, its ultimate use is in the establishment, evaluation and revision of air quality standards. This is done to some extent for some pollutants now, but several problems must be solved before plant biomonitoring data can be used more widely in relation to air quality standards.

A. Relationship of Plant Responses to Ambient Pollutant Concentrations

There is a need to greatly increase accuracy and reliability in the quantitative assessment of plant biomonitor responses to specific pollutants. Quantitative chemical analyses of plant tissues for sulfate, HF and heavy metals are the most reliable reflections of pollutant concentrations and uptake. Ozone, PAN and ethylene, however, do not accumulate in plant tissues and chemical analyses cannot be used. Evaluation methods for biomonitor plants must be developed that are quantitative in terms of the magnitude of the plant response and directly related to ambient air concentrations of a specific pollutant. It is known, for example, that a linear dose/response curve for O_3 and Bel-W3 tobacco can only be obtained when enough plants are used in a grid pattern that is both large enough and in a configuration designed to accurately biomonitor the O_3 dose in relation to prevailing winds, topography, etc.

B. Relationship of Plant Responses to Human Health

Many plants are known to respond quickly to low concentrations of air pollutants in predictable ways. As a result, plants are generally thought to be more sensitive to air pollutants than are animals and humans. Plants make excellent experimental organisms, as they are easily propagated and can be sacrificed in large numbers to obtain cumulative information on a large scale. We know more about pollutant effects on plants than on animals and humans. It is our rather sketchy knowledge of pollutant effects on humans, however, that is usually used in determining air quality standards, not the data on pollutant effects on plants. An example of this is the recent decision by the US Environmental Protection Agency to increase the O_3 standard by 50% from 0·08 to 0·12 ppm. Only fragmentary evidence on human health effects was used to establish this new standard. There is a very urgent need to establish a relationship between plant responses to pollutants and responses in humans to the same pollutants at the same concentrations.

SELECTED BIBLIOGRAPHY

1. ANON. (1979). *Cleaning the air, EPA's program for air pollution control.* EPA Pub. 48/8. 16 pp.
2. BOUHUYS, A., G. J. BECK and J. B. SCHOENBERG. (1978). Do present levels of air pollution outdoors affect respiratory health? *Nature (London)*, **276**: 466–71.
3. ENGLE, R. L. and W. H. GABELMAN. (1966). Inheritance and mechanism for resistance to ozone damage in onion, *Allium cepa* L., *Proc. Amer. Soc. Hort. Sci.*, **89**: 423–30.
4. FEDER, W. A. (1978). Plants as biossay systems for monitoring atmospheric pollutants. *Environ. Health Perspectives*, **27**: 139–47.
5. FEDER, W. A. and W. J. MANNING. (1979). Living plants as indicators and monitors. In: *Handbook of Methodology for the Assessment of Air Pollution Effects on Vegetation.* (Heck W. W., S. V. Krupa and S. N. Linzon (eds.)), Air Poll. Control Assoc., Pittsburgh, Pa. pp. 9–1 to 9–14.

6. GABELMAN, W. H. (1970). Alleviating the effects of polution by modifying the plant. *HortScience*, **5**: 250–2.
7. HECK, W. W. and A. S. HEAGLE. (1970). Measurement of photochemical air pollution with a sensitive monitoring plant. *J. Air Poll. Control Assoc.*, **20**: 97–99.
8. HECK, W. W., A. S. HEAGLE and E. B. COWLING. (1977). Air pollution: Impact on plants. In: *Proc. 32nd. Meeting Soil Conservation Soc. of Am.*, pp. 193–203.
9. JACOBSON J. S. and A. C. HILL. (1970). *Recognition of Air Pollution Injury to Vegetation: A Pictorial Atlas.* Air Poll. Control Assoc., Pittsburgh, Pa.
10. JACOBSON, J. S. and W. A. FEDER. (1974). *A regional network for environmental monitoring: Atmospheric oxidant concentrations and foliar injury to tobacco indicator plants in the Eastern United States*, Bulletin No. 604 of the Massachusetts Agricultural Experiment Station, University of Massachusetts, Amherst, 31 pp.
11. MANNING, W. H. (1980). The ozone standard. *Garden, the Journal of the Botanical Garden Society*, **4**: 2–3, 31.
12. MANNING, W. J. and J. F. MCCARTHY. (1976). Indicator plants for detection of atmospheric oxidants in Massachusetts: Alternatives to Bel-W3 tobacco. *Proc. Am. Phytopathol. Soc.*, **3**: 308.
13. MANNING, W. J. and R. L. GILBERTSON. (1980). Grapevines as perennial bioindicators of oxidant air pollution. *Phytopathology*, **70**: (In Press).

Index